Other books by Gerald M. Lemole, MD

After Cancer Care
(with Pallav K. Mehta, MD, and Dwight L. McKee, MD)

The Healing Diet

Facing Facial Pain
(with Emily Jane Lemole, MA)

LYMPH & LONGEVITY

The Untapped Secret to Health

GERALD M. LEMOLE, MD

with Sandra McLanahan, MD,

Dwight McKee, MD, and Ted Spiker

Illustrations by Nadia Chen

SCRIBNER

New York   London   Toronto   Sydney   New Delhi

This publication contains the opinions and ideas of its author. It is intended to provide helpful and informative material on the subjects addressed in the publication. It is sold with the
understanding that the author and publisher are not engaged in rendering medical, health,
or any other kind of personal professional services in the book. The reader should consult
his or her medical, health, or other competent professional before adopting any of the sug-
gestions in this book or drawing inferences from it.

The author and publisher specifically disclaim all responsibility for any liability, loss or
risk, personal or otherwise, which is incurred as a consequence, directly or indirectly, of the
use and application of any of the contents of this book.

Scribner
An Imprint of Simon & Schuster, Inc.
1230 Avenue of the Americas
New York, NY 10020

First Scribner hardcover edition October 2021

SCRIBNER and design are registered trademarks of The Gale Group, Inc.,
used under license by Simon & Schuster, Inc., the publisher of this work.

For information about special discounts for bulk purchases,
please contact Simon & Schuster Special Sales at 1-866-506-1949
or business@simonandschuster.com.

The Simon & Schuster Speakers Bureau can bring authors to your live event.
For more information or to book an event, contact the Simon & Schuster Speakers Bureau
at 1-866-248-3049 or visit our website at www.simonspeakers.com.

Manufactured in the United States of America

1   3   5   7   9   10   8   6   4   2

Library of Congress Cataloging-in-Publication Data has been applied for.

ISBN 978-1-9821-8025-6
ISBN 978-1-9821-8027-0 (ebook)

To John Daly, MD (1947–2021),
Dean of Temple University School of Medicine,
taken from us suddenly and unexpectedly on March 26, 2021.
A brilliant dean, teacher, scholar, and administrator,
and a superb surgeon and researcher, but most important,
a true friend and noble human being

Contents

Foreword   xi

Introduction: Get to Know Flow   xiii

  1: Flow on the Go   1

  2: Heart Disease   13

  3: Cancer   29

  4: GI Disorders   41

  5: Weight Management   55

  6: Brain and Mind Conditions   63

  7: Fuel for Flow   83

  8: Fast Flow   101

  9: Smooth Flow   117

10: Well Meaning   131

Appendix

    A Flow-Friendly Diet   140

    Recipes   143

    Basic Yoga for Lymph Class   181

Acknowledgments   197

Index   199

Foreword

While Western medicine often thinks of the body solely in terms of its parts—the weakened heart, the aging brain, the arthritic knee—the reality is that the body isn't about the parts. It's about the whole.

This whole is made up of systems that work together—sometimes in concert, sometimes not. And for optimum health, wellness, and youthful energy, all of those bodily interactions *should* work together and do have influence over everything that we do.

In *Lymph & Longevity*, Dr. Gerald Lemole explores a system that truly has been untapped. Most of us hear the word "lymph" and associate it with the word "cancer." But as Dr. Lemole explains, the lymph system is a majestic and crucial one, as it's responsible for keeping our many systems working properly, efficiently, and optimally.

Dr. Lemole, a pioneering cardiothoracic surgeon and one of the first doctors to see how lymph influences all aspects of our health, explains that the lymphatic system is a "secret river of health." And it's true. The lymphatic system—made up of nodes, vessels, and fluids—has not been well studied or well understood in mainstream and medical circles for a variety of reasons. For one, we haven't figured out how to measure it—to quantify whether you have good flow or low flow. But as Dr. Lemole spells out, the role of lymph works by protecting organs, blood vessels, and systems from cellular trash, toxins, and other threats to the way we function.

The crucial point: Have a strong lymphatic flow, and live healthier.

Have a stagnant or slow flow, and you're at higher risk for disease and dysfunction.

In this book, Dr. Lemole will take you through the biology of this system to teach you about its significance and its influence on the whole body. He'll do this by showing you:

- The importance of the lymphatic system as its own entity and not just as a supporting cast member to such star performers as the circulatory and digestive systems
- How it plays a role in the management of many conditions associated with aging, such as diseases of the brain, heart, and immune system
- How to use lifestyle techniques to "improve your flow" to help quiet chronic inflammation, which lies at the heart of many conditions associated with aging, low energy, and overall dysfunction

Dr. Lemole shares my passion (backed by scores and scores of research) involving food and its power to heal. In this book, he outlines the foods and associated nutrients that have been shown to influence the lymphatic system. In addition, he shares some cool and fun ways to improve the function of your lymphatic system (who would have guessed that a good joke could actually turn on the faucets for better flow?).

The story of the lymphatic system is a fascinating one—one that hasn't been told in the context of overall health. And Dr. Lemole, who has spent his entire career as a pioneer, is at it again: teaching us about a system that hasn't been well understood—but should be.

—Mark Hyman, MD, Pritzker Foundation
Chair in Functional Medicine, Cleveland
Clinic Lerner College of Medicine

INTRODUCTION

Get to Know Flow

Anyone who has seen a picture of a skeleton or looked at a poster of the human body on a doctor's office wall has a basic understanding of how the body is structured: We've got a lot of hard stuff (our bones), a lot of soft stuff (our organs), and a bunch of tunnels and highways that carry substances from point A to point B. Our blood travels in and out of the heart, our thoughts and movements travel from one neuron to another, and our food travels in through one orifice and out another.

While you can make a case that most people don't need to have a PhD or MD to live a healthy life, it turns out that understanding the systems of the body is important. In fact, research has shown that one system in particular—the lymphatic system—is the key factor in longevity, disease prevention, and living a healthy and vital life. Yet nobody—not the research community, not the medical community, certainly not the public—knows much about it.

I want to change that, because after decades in the medical community as a cardiovascular surgeon, I have seen the effect and the promise of a healthy lymphatic system. And I see what an impact it can have—for better or worse—on your health, wellness, and longevity.

In *Lymph & Longevity*, I have teamed up with two of my most trusted and valued colleagues in integrative medicine—Dr. Dwight McKee and Dr. Sandra McLanahan—to explain the power and potential of the lymphatic system—something I often refer to as "the secret river of health."

This secret river, as the name implies, operates virtually unnoticed. "Lymphatics" is a word often associated with cancer, as this is one of the major channels by which cancer can spread within the body. Researchers hoping to find a magic bullet to stop the spread of cancer via the lymphatics has meant that few have looked at the physiology of the normal system of lymphatics—and how it can influence other areas of the body beyond the way in which it works in cancer.

The lymphatics are where our daily biological battles are won and lost. Research is beginning to show that a healthy lymphatic system has a role in preventing and decreasing not only cancer, but also heart disease, brain problems, gastrointestinal issues, and much more.

Lymph has a hand in virtually every major problem that can happen in the body. Because of this, it's actually been called the "Cinderella of medicine"—underappreciated, but doing all the important work.

Containing tissues, organs, fluid, and vessels, the lymphatic system touches every other system in the body. The simple fact is, the better the lymph runs, the better the body runs.

Early in my medical career, I was doing heart surgery. Working with a senior pathologist, I saw that the lymphatics of people suffering from cardiac problems were cloggy, foggy, and polluted. The lymphatics in healthy patients? Crystal clear. It was the difference between a sewage system and Caribbean waters. That led me to decades of research and clinical work looking at the role of the lymphatic system in both health and disease.

Using the metaphor of this river that runs through the body, we hope to take you on an anatomical journey that will demystify the way lymph works so that it becomes as common a biological concept as cholesterol or blood sugar.

Our goal is to help you improve your lymphatic flow so that your body is better able to clear out toxins and waste products and clean up the domino effects of dysfunction that cause injury, inflammation, and disease, which ultimately results in chronic poor health. You will learn how lymph plays a direct role in various systems of the body (cardiovascular, neurological, immune, and more), and you will be given a plan of action that will help you fortify and facilitate your own lymphatic flow.

The human body is somewhat like a walking aquarium; that's why we can safely walk on land: we brought the ocean with us. Our bodily fluids, both blood and lymphatic, are similar in composition to seawater: salty, filled with nutrients that make up our own personal ocean. In an aquarium, keeping the water clean and flowing is essential to the health of whatever is living in it. In this case, our cells are fed oxygen and other essentials by the blood, and their health is maintained by circulation of lymph. The white cells of the lymphatics accomplish repair work, and the waste products of the cells are removed by being bathed in the oceanic elements of the lymph system.

The lymphatic system—which runs between your blood vessels to clear toxins from the body and help deliver nutrients—operates unlike any other system of fluid in your body. It works without a central pump (like the heart), it does not have clear diagnostic markers to measure its health or function, and, as noted, the clinical medical community has spent little time looking at its role.

No doctor comes into a checkup and says, "How's your lymphatic flow?" Sure, a doc may feel for swollen lymph nodes to detect an infection, but the reality is that the lymphatic system is a medical mystery to most people, patients and physicians alike. Yet it holds the key to your health, your longevity, your ability to fight disease, your energy, and your wellness.

Imagine a set of secret rivers hidden deep in the world—somewhere far away from the hustle and bustle of civilization. They've never been explored. Few people know they exist. One of those rivers is the most magical, pristine, and awe-inspiring body of water. The other is polluted, dirty, and stagnant. These are the ways our lymph systems can work, and this book is going to take you on a journey to discover your own river of health.

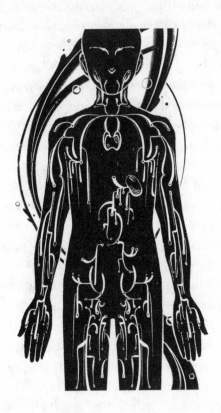

Because when you understand your own river of health—and can act upon things that will improve its flow and function—well, it's all smooth sailing from there.

Still understudied and virtually never measured in daily out-patient or even hospital-medicine practice, the lymphatic system will turn out to be the biggest lightbulb area for health—as a clear, clean, and strong-flowing lymphatic system will prove to be one of the major mechanisms for better health.

Understanding lymph is vital: When you have a tangible understanding about how the body works, you are better able to put healthy behaviors into action. You can visualize the way the body is functioning and learn how to support health, rather than being unaware that you may be living with a stagnant and polluted lymphatic system, which will eventually result in poor functioning of multiple systems within your body.

Given the fact that we're in a global health crisis—in terms of obesity, cancer rates, heart disease, diabetes, and immune problems—this book can help change the way the world and the medical profession think about the lymphatic system. Using pioneering research (in which I took part when groundbreaking data were revealed about the lymphatics in a 1981 article published in the *Annals of Thoracic Surgery*), we hope to teach you how to leverage your lymphatics to live healthy, strong, and long.

This is the healing power of flow.

Why is the lymphatic system so critical? Because when it's not running smoothly, you're at greater risk of developing chronic degenerative diseases that decrease your quality of life, increase your healthcare costs, and increase your chances of premature death. But a river that runs well? It's like fountain-of-youth fluid scrubbing your whole body. Here's how it works:

- One of the signature traits of chronic degenerative disease is chronic inflammation. This happens when the immune system becomes dysfunctional and your body is being overwhelmed by bacteria, free radicals, and other substances that cause disturbances in the way your body works—putting your body in a state of high alert all the time.
- Why are lymphatics so important? One, they deliver information about those toxins in your body to the immune

cells so they can destroy or neutralize harmful substances. Two, the lymphatic system carries away harmful substances and any by-products formed by them. Three, they aid in the cleanup process to help heal and rebuild molecules in the area. Without effective and efficient lymphatic flow, any of those three processes are compromised—putting you at risk of chronic inflammation and chronic carriage of toxic substances, which interfere with healthy metabolism and bodily function.

- This process has to happen quickly: get the message out, destroy attackers, clean the area. The lymphatic transport system—the secret river of health—is the pathway that cells must take to do these three jobs. Any delays contribute to chronic inflammation or autoimmune disease.

It's not a simple process, and it involves many parts of lymphatic fluid—its flow, viscosity, composition, and volume. But it is a process that you will understand, as we take you through the various parts of lymph and how it works in context of your organs and systems of the body.

Best of all, you can do things that help improve your lymph flow. We will talk about these strategies throughout the book.

This ability to improve your health has really emerged, in part, because of the field of epigenetics. It used to be believed that we had little control over our genetic destiny—that our genes are our genes, and there's little we can do about the way they function. The old thinking was that you were a prisoner of the genes you were handed.

But things have changed.

The emergence of epigenetics has taught us that things we do or don't do have the ability to change the way genes are expressed—an on or off switch of sorts. And that has major implications about the way our bodies function.

The history of epigenetics is a fascinating one. Before the decoding of the genome by Francis Collins around the turn of the century, it was believed that humans had about 150,000 genes based on the different proteins, amino acids, fats, and carbohydrates that we make daily.

Much to the surprise of geneticists, the genome was found to contain 25,000 genes, of which 90 percent were considered "junk genes." This meant that they would have multiple functions and work with other areas of the genome. Why is that important? The way genes were expressed could be influenced by messengers to open or close the gene.

This meant that your genes weren't a strict code of dictating how you were going to work, but rather a fluid file that could change and help make you healthy or not by helping determine which genes would be suppressed and which would be expressed. The implications for what you can do with cancer-promoting genes or obesity-related genes are astounding.

The things that we care about in terms of lymph are quality and flow. You might think those things are predetermined by our genetics. But we do have the ability to help improve lymph—and our overall health in the process.

Improving your lymphatic flow relates to the foods you eat, how you exercise and relax, saunas, laughter, and even spirituality. We'll give you ideas and action steps to get your lymphatic system in working order using three basic methods: feeding your flow with good nutrition, creating fast flow with intense exercise, and maintaining smooth and unobstructed flow with meditation and relaxation.

Today, the world uses the word "flow" to define how we feel mentally—workflow, sports flow, go with the flow.

But in this book, we want to introduce you to a whole new concept of flow—one that will change your mind, change your body, and change your life.

LYMPH
&
LONGEVITY

CHAPTER 1

Flow on the Go

Understanding How the Secret River Runs
Will Help You Realize the Power and Potential
of Your Lymphatic System

For some, "flow" is a state of mind. For others, "flow" can describe the way you dance. Or maybe a chart, an influx of cash, or the pouring of a round of drinks.

When it comes to the human body, "flow" is, well, just about everything. Think about all the flow-based systems that are required to make you *you*. Blood flows through your blood vessels. Your meals flow from your mouth to your anatomical exit ramp. And even your brain operates with an electrochemical flow of neurons and other elements darting and swirling inside that skull of yours.

Here is the amazing thing about your lymphatic system: Besides being its own flow system—made up of lymph cells and fluid—it also intersects with all the other flow systems in your body. That's what makes it such an important—albeit understudied—river of health.

So when you think about the lymphatic system, as well as other flow channels in the body, it does help to think about the health of a river and its tributaries. The health of a body of water depends on its overall cleanliness (clear, with few toxins), and its power lies in its ability to move strongly and swiftly.

The lymphatic system needs to work in this way as well: clean, strong, free of toxins. When it works like that, it can do its job to keep *you* healthy. But when it's murky, stagnant, and polluted, it won't function well.

Here's the big thing to keep in mind: Unlike the circulatory system, the lymphatic system doesn't have markers and symptoms and biological billboards to signal that something is wrong. You can't diagnose a blockage with imaging techniques. You can't get a reading with cholesterol levels. And you can't unblock a pathway with a bypass procedure.

Your lymph system is secret, because—while we know it's there and what role it plays—everyday folks and our medical practitioners have very little ability to really understand what is going on in individuals. And that—we believe—is all the more reason why this secret river needs to be more like an anatomical attraction. So thinking about it, paying attention to it, respecting it, and focusing on it can play a role in your overall health.

Also, the lymphatic system relates to all the other systems in the body—immune, circulatory, endocrine, nervous, and so on—so we can't talk about overall health and wellness without thinking about all the touch points that lymph has *everywhere*.

Throughout this book, we will take you through some of the most pressing diseases, conditions, and worries that a person can have—and show you what kind of role the lymphatics play. First, though, it will help you to understand the big picture with an up-close look at the river: how it works, what it does, why it's tricky; we'll examine what potential and power it has when it comes to improving your wellness, increasing your longevity, and decreasing your risk of developing disease.

Your Lymphatic System Is Your Body's
Maintenance Department

Why is that? Because its job is to optimize the health of all of your cells. It works a lot like the cardiovascular system in that it's made up of a series of small pipes that move fluid throughout the body. These vessels are much smaller than your blood vessels, and they carry a clear, watery fluid called lymph.

Interstitial fluid (the fluid between cells that is not technically considered lymph) enters lymph capillaries, and this fluid goes to larger lymphatic vessels through the lymph nodes. Lymph fluid works by conveying substances (large proteins such as antibodies, enzymes, and cells of the immune system) that nourish, protect, and coordinate the body, as well as carrying out the cells' waste (oxidized lipids, damaged proteins, large-molecule toxins) via its collecting tubes.

Lymph, when working well, traps the threats to your body (such as toxins and viruses) and gets them to the lymph nodes, where they can be neutralized and processed to be removed from your body.

As you might imagine, having a better flow of good things moving around the body and bad things moving out is crucial to all your body's workings. In addition, the thickness (viscosity) of lymph fluid is affected by water and other compounds in the fluid, especially from your dietary choices, and in turn influences the functioning of all the systems of your body. The message being, you have some control over how well your lymph runs, just as you have some control over the quality of your blood flow through your blood vessels and cardiovascular system.

There Is No Central Pumping Mechanism
to Power Your Flow

Unlike your cardiovascular system, which operates because blood flow starts with the pumping action of the heart, your lymph system has no central hub to kick-start the flow of fluid. Instead, the flow is orchestrated by pressure from various muscular systems and the nearby pulses of the vascular system, which often run right next to the lymphatic vessels, all of which help to pump lymphatic fluid around your body. But here's the really interesting thing: You can control some of this flow through your own actions, such as deep breathing and muscular movements. Your diaphragm, the large muscle between your abdomen and chest cavity that primarily moves air in and out of the lungs, functions as a sort of pump for the lymphatic system—especially when it's moving deeply, or in a coordinated fashion, as with yoga breathing exercises.

Your Lymph Has Three Main Functions

Your body is thought to have some 37 *trillion* cells. Every single one of them has intimate contact with lymph—that is, other systems in your body rely on lymph to function. That's because this river of health is the channel through which messages travel to and from the body's systems with one goal in mind: getting your body into a state of homeostasis (that's a medical term for healthy, balanced function). It does this with the overarching goal of protecting your body from the damage of toxins, and it does this through three main functions:

- Transporting toxins, immune cells, and messages throughout the body via the lymphatic fluid.

4

- Transporting larger endogenous (made by your body) amino acids, proteins, and fatty substances from your gastrointestinal system to your liver.
- Storing immune cells and transporting immune signals. Within your lymph system, you have five hundred to six hundred lymph nodes, where white immune cells wait to help your body's defenses not only against harmful bacteria, viruses, yeasts, and other obnoxious bugs, but also to remove aging, damaged, and cancerous cells. Lymphatic vessel lining cells can receive and transmit messages directly or through the nervous system and can respond to chemical substances like cytokines (immune substances secreted by the cells) to increase or decrease lymphatic flow.

Your Lymphatics Have a Unique Way of Working

We don't want to overwhelm you with a lengthy anatomy lesson, but we do think it's important to have some basic knowledge about how the lymph system operates so that you have a more visceral understanding of the mechanics. That, we believe, will help you as you develop new habits for helping to improve your flow. When you understand the "why" behind the biology, it makes the "what to do" much clearer and more powerful. So let's take a look at the mechanics and biology of how flow works.

The lymphatic channels have a muscular layer that can create flow by dilating or constricting. Lymphatic vessels can become leaky and exude fats like triglycerides and proteins like gamma globulins (antibodies) that are necessary for a healthy immune system. These substances can then build up around the blood vessels located in the abdominal region (called mesenteric vessels) that carry essential nutrients and messengers from the gastrointestinal (GI) system, and thereby create serious metabolic disorders. The fluid located within the lymph tubes, while still surrounding the tissues, can be very watery and flow easily or become sluggish and sticky in situations where there is poor water intake or from the effects of toxins and diet, which act to diminish flow.

As oxygen and nutrients are circulated throughout the body, the serum of the blood trans-

> ## Things That Influence Lymph Flow
>
> **Fluidity:** The overall ability to flow
> **Muscle Action:** Smooth muscles that allow vessels to pump fluid
> **Peristalsis:** Muscle contractions that help fluid flow
> **Contraction/Dilation:** Tightening and opening of the lymph vessels
> **Pressure Gradient:** Lymph tends to move downstream
> **Sclerosis:** Scarring of lymph tissue

ports them through the first part (proximal) of the capillaries; then at the end (distal) capillary section, metabolites, small-molecule toxins, and carbon dioxide reenter the blood circulation. Any fluids that leak out of the capillaries create our internal ocean of lymph, bathing our cells in fluids. Lymph vessels collect some of this fluid, then take it back into the blood.

A significant amount of fluid, 10 to 15 percent, remains in the tissue (if more fluid than this remains, the buildup is known as edema). The lymphatic system carries these fluids back and delivers them into the large veins in the chest, where they are then recirculated and cleared.

In addition, lymphatic endothelial cells also play a role in sending and receiving signals (via nerve impulses and other messaging). These work by regulating the diameter of the vessels to help with tissue clearance and immune response.

Since there is no pump to move the lymph forward, like the heart does with your blood supply, lymphatic flow depends on several components: lymphatic pulsation, spasm or flaccidity, pulsation of nearby arteries, muscular activity in the area, negative pressure in the chest, lymph viscosity, and gravity. (Negative pressure means that the pressure between the lungs and the chest wall is below atmospheric pressure and actually increases in negativity during a deep breath.)

The remarkable thing is that you have the power to increase lymphatic flow by the positive health actions you take. For example, some foods and nutrients can increase your lymphatic clearance, and some physical activities can do the same.

Your Lymphatics Are Prepared to Fight Against Disease

It's important to know why good flow is necessary: It helps clear toxins and damaging materials from parts of your body (all tissues contain lymphatic channels) and allows immune-cell responders to travel to damaged and inflammatory sites. Good lymph flow signals that help is needed quickly and these responders can get there fast.

In other words, since the lymphatics are responsible for acting as conduits for immune system molecules and cells that both initiate and clean up inflammatory tissue, the lymphatics can enormously influence the initiation and outcomes of chronic degenerative disease. This holds true for not only diseases associated with immunity (like arthritis), but also has implication in the hardening of arteries and some neurological disorders. Back in the early 1980s, when I first described the involvement of the lymph system in arteriosclerosis, knowledge of the immune system was limited, but new technology and techniques have shown how immunity, the lymphatics, and chronic disease are related.

This typically involves an inflammatory process and a disturbance in lymphatic flow. How?

Three ways, primarily:

1. Decreased functionality of a system. In the cardiovascular system, for example, lymph could have an effect on the delivery of nutrients and oxygen to essential organs of the heart and blood vessels, and this can contribute to the hardening of your arteries and/or clearance of them.

> ## Five for Flow: Supplements for Lymphatic Drainage
>
> - Diosmin
> - Pycnogenol
> - Horse chestnut
> - Nattokinase
> - Polyphenols
>
> (See more about supplements in the Appendix, page 177.)

2. Change in the expression of your DNA. If specific signals are sent to your cells' nuclei, certain areas of the DNA may be opened or closed, activating substances such as oncogenes (genes that can cause cancer) or the release of other chemical messengers called cytokines that cause increased inflammation.

3. Leakage of vital nutrients. Inflammatory molecules carried through lymphatic channels can directly affect your cell membranes, causing leakage of vital nutrients or an attack on the proteins in the cell's nucleus, causing derangements that may lead to cancer. Improving lymphatics would facilitate early destruction of toxic substances and then prompt cleanup and minimization of tissue damage.

Lymphatic Dysfunction Plays a Role
in Three Major Systems

The first sign that something is going wrong with your lymph? Poor function of the organ it is serving. (This is called lymphatic derangement.) The three big places this takes place are:

Cardiovascular System: Your blood supplies your tissues with oxygen and nutrients. Under healthy conditions, any LDL (the potentially unhealthy form of cholesterol) that makes its way into the arterial wall is rapidly picked up by the HDL cholesterol (the good kind) and macrophages (large white cells of the immune system). These are then carried through the vessel wall to the lymphatics, which carry them back to the liver.

In the diseased state, the oxidized LDL becomes embedded in the subintimal layer (just below the cells lining the arteries) of the larger and medium-size arteries and because of poor clearance due to inflammation and decreased lymphatic transport, the cholesterol becomes built up and eventually blocks the arterial wall. To prevent this, it is essential to rapidly contain tissue damage and clear the toxins, cytokines (substances that affect immune cells), and metabolites from the area. Such key cleaning action is accomplished by your lymphatic vessels.

During my years of performing heart surgery, it became apparent to me that the lymphatic vessels on the surface of the heart can become sclerotic or scarred. This can occur with stress, smoking, lack of exercise, or poor dietary choices. Biopsies of the lymphatic vessels in effected patients made it clear that the scarred tissue prevented the inflammatory products from being removed from the affected artery.

Neurological System: It's only been recently recognized that the brain itself needs lymphatic clearance, just like other organs in the body. We never thought that the closed space contained in the skull where the brain is housed could tolerate flow of fluid in the brain tissue. Investigators at the University of Rochester discovered what they called the glymphatic system, because it involved lymphatics. They also found that the glial cells of the brain had to shrink in size to accommodate the flow of cerebral fluid into the brain. Much of this flow occurs during sleep; this is a type of "brainwashing" that is good for you!

The connection between cerebrospinal fluid flushing and the lymphatic vessels was then demonstrated by researchers at the University of Virginia, who showed a connection between the glymphatic system and the lymph channels of the dura mater (connective tissue) of the skull that subsequently connect to the deep cervical lymph system. This research helps explain how toxins and inflammatory metabolites are cleared from the depths of the brain.

Gastrointestinal System: The lymphatics of the GI system play a dual role, not only clearing toxins from the tissues but also transporting the fats, larger amino acids, and proteins from the foods you eat, moving them from your gut to your liver to be metabolized. Leaky lymphatic channels, poor lymphatic communication between the immune, nervous, and endocrine systems can create a situation where gamma globulins, cytokines, triglycerides, and other fats exude out of the lymphatic system into the peripheral tissues, creating pressure on the vascular bed.

This creates a large belly and prevents good messengers coming from the gut from fulfilling their mission. Abdominal obesity can be a harbinger of metabolic syndrome and predictor of hypertension, diabetes, and other illnesses. In general, then, excess belly fat prevents toxins from leaving your body, healthful nutrients from entering and circulating, and thus your secret river of healthful lymph from doing its wondrous work.

We Do Not Know How to Measure Your Flow

Want to know the health of your heart? We can take some readings—blood pressure, heart rate, cholesterol. We also have markers for inflammation, indicators for blood sugar, and measurements about whether you're vitamin deficient; we can even image the blood vessels in the heart and, with similar imaging techniques, detect blockages in flow throughout the body's entire arterial system. But want to know how well your lymphatics are flowing? Good luck.

That's one of our big roadblocks in the area of lymph. We do not have an efficient and accepted way to measure its content and function. That makes it clinically difficult to discuss: How can we advise people that the key to good health is to improve flow if we have no way of measuring its current health and whether or not you have been able to improve it?

Herbs That Support Lymph

Siberian ginseng (*Eleutherococcus senticosus*) has, among its many actions, the ability to stabilize lymphatic vessels by protecting and enhancing the endothelial cells (the ones on the lining) of the lymph system. The use of the herb has been shown in clinical trials to stimulate lymph drainage to such an extent that edema of the lower limbs was significantly lowered two and four hours after taking it. Another supplement that seems to help with protecting the various parts of the lymph system is Japanese knotweed (*Polygonum cuspidatum*), a rich source of resveratrol. Additionally, both pleurisy root (*Asclepias tuberosa*) and inmortal (*Asclepias asperula*) can help stimulate lymph drainage from the lungs. All these can be found on Amazon or in health food stores.

CHAPTER 2

Heart Disease

With Good Flow, You Can Prevent Clogs, Blockages, and All Kinds of Damage that Contribute to Heart Disease

The heart may very well be the most iconic emoji of our time, and it's certainly been a spot-on word for country music lyrics, a convenient shape for boxes of Valentine's candy, and the inspiration for a tattoo or two. But in my world, the heart is not only a symbol for many things in life. It *is* life. And death.

Heart disease, as you know, is the number one killer in the United States, so it's no surprise that the medical attention spent on the heart is enormous—everything from clinical research and doctor directives to surgical advances and pharmaceutical interventions.

As someone who has spent his entire career in the cardiovascular field (I was on the first successful heart-transplant surgical team back in 1968), I have lived a dictionary's worth of verbs when it comes to the heart. Touched it. Studied it. Fixed it. Watched it. Transplanted it. Stopped it. Started it. Thought about it. Admired it.

So maybe it goes without saying that everyone knows that the heart is our central organ of life. But perhaps everyone doesn't know that what we've focused our attention on for the last several decades is only part of the puzzle that will keep our hearts beating for a very long time. What do I mean? In my mind, in order to do a better job of managing heart disease (to increase both longevity and quality of life), we need to make a dramatic shift in two key areas:

Fixing the problem is great. Preventing the problem is greater. The medical community has become incredibly efficient and effective in surgically treating people with life-threatening heart problems. The evidence: 98 percent of surgical patients survive, and 92 percent in emergency cases survive. (Heart attacks have been found to be 11 percent higher on Mondays, by the way; the risk is lowest on Saturdays, for reasons that haven't been confirmed. Popular theory would be that the typical workweek flow has something to do with it.) We have developed many different ways to bypass blockages and ensure that blood can move freely to and from the heart. We have developed medications to lower high blood pressure and to reduce high cholesterol. We have surgical advances that can repair damaged hearts in the most minimally invasive ways. In all, we have unprecedented methods for *reacting* to our most pervasive problem.

But what if we flipped the narrative so that the majority of members in our medical community weren't just fixers, sewers, and drug dealers? What if we were problem-solvers before the problem even emerged? Yes, there's a lot of chatter about things you can do to live a healthy lifestyle, but the common reality is that many of our medical professions and educational systems focus on therapy of diseases rather than stopping them before they develop. The truth is that medical school spends little time—if any—talking about the biological basis for prevention. Minimally invasive education, perhaps?

If we're going to help reduce the risk of heart disease, the focus can't be on the fixing. It has to be on the upkeep and maintenance. Wouldn't you rather have a car that didn't break down instead of one that needed to be repaired? A phone that didn't fritz out? Pipes in your house that never corroded?

In every aspect of life, proper and consistent maintenance—and a strong infrastructure—is preferable to a catastrophic problem that needs a heroic repair. That's the same approach we need to take for your heart.

We haven't even tapped into the underlying mechanism for preventing heart disease. My guess is that if I said the phrase "heart disease" or "heart attack," you could rattle off a number of causes right off the top of your head. Clogged arteries. High blood pressure. High cholesterol. Damaged valves. Plaque buildup. And you'd be right. The heart and cardiovascular systems are a complex part of your anatomy with a combination of biological, chemical, and electrical systems that make it all work. Any number of root causes can contribute to the malfunctioning of the heart (and thus the necessitation to have some of the interventions I mentioned earlier).

Yet the central theme of this book is that the lymphatic system is one of the foundations for overall health and the health of a variety of organs. Strong lymphatics mean that you can have a strong heart—mainly, as this chapter will set out to explain, by playing a role in any of the number of causes of heart disease. Almost forty years ago, I authored a paper with the hypothesis that poor lymphatic flow contributed to the start of heart disease. Today, as I look at the growing body of evidence and research, I see it more clearly. Because cardiovascular disease is actually a disorder caused by inflammation and damage caused by substances related to that inflammation, the lymphatic system—and its role in clearing toxins and quieting inflammation—is the secret key to improving the long-term health of the heart. And the lifestyle choices you make play a role not only in the areas like blood sugar, cholesterol, and more, but also in the areas of lymphatic flow, which serves as the central controller of these factors.

In this chapter, I will outline how the heart works, how it can become damaged, and then how the lymphatics are structured to help prevent long-term problems. Now, I don't expect that the lymphatics will take their rightful place in love-song lyrics, but I suspect that you will start seeing that if you want to keep your heart beating, you must keep your lymphatics flowing.

The Anatomy of Heart Health

Chances are, you already know the basic function of the heart: The bass drum of your body, your heart is the muscular organ that sets the beat and rhythm of all bodily functions by generating the pressure to move blood throughout your body. Since blood is what carries oxygen and nutrients to your organs, your body can't function without your heart working like a well-bloodied machine.

For these purposes, we don't need to go into the full-fledged inspection of the structure of the heart. A quick overview: The heart has four chambers, valves, and an electrical system that controls heartbeat. Capillaries are small blood vessels where oxygen, nutrients, and other substances travel between your blood and parts of your body. Veins move blood to your heart and arteries move blood away from your heart.

In a healthy circulatory system, the heart pumps blood through your various blood vessels. This fuels your organs—your brain, your lungs, the ones that make up your digestive system, etc.—with the nutrients they need to function. Both the rate of the blood flow and the substances in the blood play a role in the functioning of the organs and how your body works. When everything is running smoothly—nutrient-rich blood delivered at an efficient rate—then all's fine. Trouble starts when something disrupts the ability of blood to go to and from the heart and get delivered throughout your body.

Here's how it happens (this will provide the foundation for why lymphatics are so important):

Your arteries are made up of three layers, and as you might suspect, they can be quite delicate. When your arteries are healthy, the inner lining (called the endothelium) is as smooth as a granite countertop, and blood can flow freely throughout the body.

Now, those delicate arterial walls are susceptible to damage as they're exposed to various things like high blood sugar, aging, cigarette smoke, high blood pressure, and nutritional toxins like free-floating sugar. These can all chip away at the arterial wall.

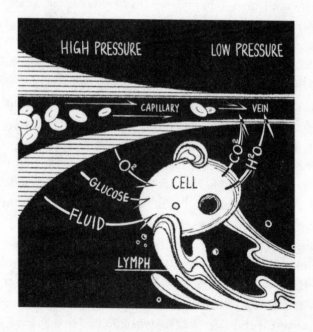

As we explained earlier, metabolites and cholesterol delivered through the capillary wall are different from those wedged in a wall of thick artery. Biologically, cholesterol is designed to *help* you. Cholesterol is manufactured by the liver and acts as a building block of the cell membrane; in fact, it's essential to ensure healthy cells. Cholesterol strengthens cell membranes, nerve sheaths, and participates in the production of hormones and bile. But it is only beneficial when it is delivered through the A-V capillaries, which are very thin.

The cholesterol that lodges itself into the spaces beneath the area where the endothelium is missing is not healthy or reparative. It's like snow falling through missing shingles in a cabin roof. If it's not shoveled away through the windows (the lymphatics), the rooms can be destroyed.

The problem occurs when cholesterol is oxidized—as is the case when it comes in to help repair arterial damage. This is what irritates the cell wall. The body views the oxidized version as an insult.

That irritation triggers an immune response; immune cells come in to try to fight the problem, which then sets off a whole chain of events that causes inflammation—a signal in your body that some kind of fight is going on.

As this process is happening, cholesterol is trying to do its job in a variety of forms. It does this hero job with help from a sidekick—a lipoprotein. Because cholesterol is a fatlike substance, it can't dissolve in water and must be delivered to the cells through the bloodstream. A lipoprotein must carry it.

These lipoproteins come in two forms: LDL (low-density lipo-proteins), which is labeled as the "bad" kind of cholesterol, and HDL (high-density lipoproteins), the "good" kind. LDL stops to repair the damage on the arterial wall, while HDL comes in and clears away the cholesterol. LDL is metabolized in the tissue, while HDL carries the cholesterol out to get metabolized in the liver. Think of the HDL as some sort of cleanup crew: While the workers get a job done, someone else comes in to make sure all the debris from the work is cleaned up. This process—when it works—is a thing of beauty.

The HDL and immune cells called macrophages that enter the area are like people shoveling out the snow through an open window—this is done via the lymphatic capillary, which is different from the beneficial function of the arterial capillary.

So why does this matter? As LDL is doing its job to repair the arterial wall, the body views this as an insult to the artery, even though the intention is to heal. So another form of cells—called foam cells—engulf the oxidized LDL to prevent it from further hanging out on the arterial walls and thus creating and setting up more of an inflammatory process. The foam cells are trying to prevent the process by engulfing the cholesterol, but when they die and disintegrate, it leaves oxidized cholesterol to create more inflammation. (They die either from old age—after about a month—or from excess substances like oxidized cholesterol.) In a healthy situation, the foam cells will travel through the arterial wall to the lymphatics in the outer wall of the artery so they can be transported to the liver to process the cholesterol which could also be picked up by the HDL and carried to the liver. Now, you can imagine that this process is slowed when the river is slowed—by inflammation, viscosity, debris, lack of physical activity that pumps the lymph, and other molecular elements that change the flow and makeup of the lymph.

When the lymphatics are compromised, the body is unable to efficiently process cholesterol and decrease inflammation. This leads to a buildup of cholesterol, calcium, and other by-products. There's your clog.

If your body can't get the debris out, your artery wall is damaged and clogged. This sets up inflammatory reactions that further put the arteries in danger.

Inflammation in short periods is a good thing, because it's a sign that your immune system is fighting foreign substances. Unfortunately, there are two phases of the inflammatory process. The first is pro-inflammation (taking care of the toxins and invaders), and the second is the cleanup and reparative stage. If this second stage is delayed or never happens, it can lead to autoimmune disease or chronic inflammation. But chronic inflammation—what happens when your arterial walls are constantly damaged—is what puts your body under constant stress, as it's unable to keep up with the steady fights happening at the cellular level.

> **Factoid**
>
> A possible game changer comes from Gothenburg University in Sweden in a paper published in 2021 in the *International Journal of Molecular Science*. This paper shows how important the lymphatic system is and how intimately involved it is with resolving chronic inflammation—not by the ordinary means of anti-inflammatories but by the natural substances resolving and protecting, which heal and restore the body's tissues.

So the effect of having problems with your circulatory system is more complex than, say, just having a clog that needs to be cleared. It's multifold—it's about preventing the damage, clearing the damage, and improving the immune response to lower inflammation. When you do that, you lessen the chance of having problems associated with the heart, such as heart attacks, heart disease, and other conditions associated with damaged arteries.

It won't surprise you that the key player in helping all of these things happen is your lymphatic system.

Lymphatic Flow to Strengthen the Heart

When we dissect the root causes of heart disease, it's clear that you can lump them into many different buckets: genetics, aging, obesity, smoking, stress, and more. But what do all those things have in common? Lymphatics influence the blood-cholesterol relationship—and ultimately, those processes of cholesterol buildup and inflammation are what restrict blood flow to and from the heart and trigger heart attacks, strokes, and chronic diseases of the heart.

How do lymphatics influence the blood-cholesterol relationship?

As you remember from the previous chapter, the lymphatics work like blood vessels; they circulate the fluids between your capillaries and your body's cells. That fluid transports nutrients to the cells. The body's toxins—oxidized cholesterol and other damaged lipids, glycated or otherwise damaged proteins, and other large molecular waste broken down in the body—also travel through the lymph system to the blood stream to be cleared by the body. Oxidized cholesterol is only deposited on the inner wall—and to be cleared, it must travel to the outer wall to the lymphatics. More oxidized cholesterol in the arterial wall means a higher toxic reaction.

If you can keep lymphatics clear and flowing, little or no cholesterol would build in the arteries, because the lymphatics would be able to remove the harmful substances from the blood vessels. That's because they remove the cellular debris from the spaces between cells, meaning that toxins have less contact with the arterial wall.

In addition, lymph also addresses the inflammation part of the equation. Remember, the process of repairing damaged arteries causes a battle in which immune cells rush in to attempt to quiet things down. That causes inflammation.

If your lymph is not flowing well, it cannot get messengers to the scene of the crime and it can't get fighter cells to stop the biological conflict. A weak lymph system works against you in two ways: to even identify a fight as it starts and to fight the fight once it gets going. But if your lymphatic system is functional and free-flowing, it can do both jobs—essentially stopping the inflammation so it doesn't lead to chronic damage.

> **Factoid**
>
> Further insight into the connection between blood flow and lymph flow: When the heart lymphatics aren't working well, this can cause the coronary artery to constrict. A Japanese study showed that tying off the coronary lymphatics in pigs created a severe spasm of the coronary arteries, which can cause arrythmia and myocardial death.

Heart disease often gets categorized as a problem associated with eating too many cheeseburgers or just having bad genes. And while it's true that nutrition and genetics do play a role, the better way to think about cardiovascular disease is as a chronic degeneration of the immune system, caused by inflammation. Your arterial structures—and the ability of blood to flow through them—is hindered because of long, drawn-out battles by the immune system, which lead to narrowed, clogged, and compromised arteries. And the ability of your immune system to repair or prevent this damage hinges largely on the functioning of your lymphatics to clear the toxins and improve immune function to prevent long-term inflammation.

Later in the book, I'll be covering all the modalities that will help you improve your cardiovascular system via the lymph.

If you want your life-thumper to keep on beating, you have to keep your river of health free-flowing.

Love Your Lymph, Help Your Heart

Thankfully, there are many things you can do to improve your arterial flow, as well as your lymphatic flow. This helps clear the gunk and inflammation that causes arterial clogs and blood flow issues. Here are some of the most important things you can do:

Jam to some tunes. Anything that helps you relax and especially breathe more deeply assists your lymph to circulate. Listening to music works; singing or playing an instrument has an even more powerful effect. Orchestra conductors live longer; they move their lymph through the chest via their arm movements. Go ahead and dance, air-conduct, and enjoy your favorite music. I played music all the time in the OR as a way to help the surgical team relax, but as it turns out, it's also helpful to lymph flow. We listened to country or easy rock 'n' roll with a beat that kept everyone relaxed and in sync.

Get a massage. Massage directly moves both blood and lymph along, getting nutrients where they need to go, and helping remove waste products. In addition, when you allow your muscles to relax, the blood and lymph vessels expand, improving flow. Massage therapy has *also* been shown to reduce pain and anxiety after heart surgery. I recommend you get a massage once a month, more often if you can. You can also do self-massage—with a damp loofah sponge before your bath or shower. Always direct your movements toward the heart, to mimic your fluid flow pattern. Even a few minutes of massage daily can give you and your circulation a big boost.

Consider osteopathic manipulation. For more than one hundred years, osteopathic physicians have been using a procedure called

23

the lymphatic pump treatment for swelling (edema) and infection. But we haven't known the mechanism by which this seems to work—until recently. How does it work? While the patient is supine, the provider stands at the head of the massage table and places their hand on the chest wall (or abdomen) as the patient takes deep breaths. Lab tests have shown that this technique mobilizes the white cells from the gut associated with certain lymphoid tissue, accompanied by a significant increase in certain lymphatic flow.

Breathe deep. Especially when faced with an angry situation, you need to breathe—not just to calm down emotionally, but to keep your lymph moving. Holding your breath impairs flow. The largest lymph vessels and main lymph channels in your body are located in your chest. Whenever you take a deep breath, it moves the lymph fluid along, and the one-way valves in these tubes keep lymph from going backward. Therefore, the deeper your breathing, the more circulation of your secret rivers of life you can get. This is especially important as far as cardiovascular health is concerned. Often, stress can cause fast and shallow breathing, which does not provide good lymph flow around your heart. Try taking time for a few minutes of deep, slow breathing morning and evening, to train yourself for optimum breathing all day long. Place one hand on your abdomen; see if you can feel it move outward as you breathe in. This moves the diaphragm in a way that provides the most space for your breath. As you breathe out, tighten your belly muscles, pushing as much breath as possible upward and outward. Repeat for a few cycles, and remember to do this several times during the day, especially whenever you feel particularly tense.

Jump. A "rebounder" is a small trampoline you can use indoors to get your lymph to flow merrily gently down its stream without much effort (and it's a great way to burn calories, according to research from NASA). Utilizing the benefit of gravity, you can jump up and down or simply bounce your feet on the trampoline, and within a short time, your fluids are moved along four times faster than walking. You can plan for a short rebounding session in the morning and evening; start with just a few minutes, then gradually increase to tolerance. Most of the lymph travels through the chest in the thoracic duct (which has one-way valves). When you jump on the rebounder, you help the fluid move up and stay there when you come down. Since the lymphatic channels are next to your arteries, increasing your pulse rate and intensity (through activity, for instance) will help improve lymph flow. In addition, according to one study from Italy, jumping decreases pain severity by 88 percent—by flushing pain-triggering inflammatory compounds through the lymph system. We recommend 10 to 20 minutes, but even five minutes or just 50 to 100 jumps can be powerful.

Enjoy the sauna. Sitting in the heat can increase heart rate, which helps the heart muscle get exercise and also moves lymph fluid where it needs to go to assist in maintenance and repair of your crucial coronary arteries. Toxins from your body are carried away with sweat. One prospective Finnish study showed that the more often subjects took a sauna, the lower their risk for heart attack and overall mortality. Saunas don't take the place of exercise, but they can be an enjoyable adjunct to a lymph-hearty lifestyle. Just make certain not to overdo it—15 to 20 minutes is enough—and rehydrate with fluids afterward. Taking a sauna even once a month was found to lower heart disease risk.

Do yoga. The gentle stretching of yoga and tai chi systematically move your blood and lymph through their circulation pathways. This has been shown to directly lower blood pressure and stress hormones, so that your lymph vessels will dilate and perform their functions more efficiently. (See page 181 for some suggested poses.)

Meditate. Your heart and lymph vessels are constantly responding to your thoughts and emotions. Meditation is the process of training the mind to become more peaceful and quiet; mindfulness is taking this ability on the road. Even a few minutes of meditation in the morning and evening can allow your lymph to flow to the more urgent needs of your cardiovascular system, for recovery and prevention. Meditation and mindfulness practice have been documented to lower stress levels, and decrease cholesterol and blood pressure directly. As you relax, you breathe more deeply: This helps especially in moving the fluids through your larger lymph vessels in the chest. Pick a focus for your meditation period that feels most uplifting to you: focus on your breath or a centering word such as "peace" or "relax," or a short prayer.

Eat for flow. Certain spices and herbs have medicinal effects on lymph flow because they contain compounds that help prevent the formation of plaque and blood clots in your arteries, but they also relax your lymph vessels to accentuate flow. They lower cholesterol, decrease blood pressure, and help keep blood sugar levels steadier. Garlic, onions, cloves, cayenne, cinnamon, oregano, basil, curry, and turmeric contain vitamins and minerals that assist in vessel maintenance and repair. Try adding them to dishes you eat on a regular basis; oregano on blueberries or oatmeal can give them a surprising pop of flavor. Other healthy dietary choices include:

- Dark leafy greens in your diet—such as spinach, arugula, kale, and Swiss chard—provide an array of vitamins and minerals needed for blood and lymph vessel repair. Their magnesium acts to dilate the vessels, while their abundant vitamin K, a fat-soluble vitamin, is known to help prevent heart disease. Dark leafy greens also contain calcium that is better absorbed than calcium found in dairy. Aim for two servings per day.

- Red and orange foods—such as pomegranates, carrots, red peppers, squash, and tomatoes—provide vitamin C, a collagen-enhancing ingredient, as well as carotenoids and flavonoids, which provide antioxidant and anti-inflammatory effects. Adequate collagen is necessary for care, maintenance, and rebuilding of vessel walls.
- Blueberries, blackberries, and dark-blue grapes contain anthocyanins, which give them their dark-blue color and act as powerful antioxidants to prevent and reverse the aging process of your vessels. You can put them in salads, and especially use them as substitutes for inflammatory desserts made from sugar or refined flour.
- Polyphenols—micronutrients found in most every plant—have a variety of health benefits. You can find them in beans, berries, veggies, and more. The reason we love them: They're known lymphagogues—meaning they promote the production of lymph.

To sum it up, a saying from the award-winning Michael Pollan, author of *The Omnivore's Dilemma*: Eat (real) food, not too much, mostly plants.

Consider supplementation. The standard American diet (SAD) that most Americans follow has way too many simple carbohydrates like sugar and starch, as well as unhealthy oils that lead to inflammatory reactions. A healthy diet should counteract the loss of the micronutrients in processed foods, but there may be need for supplementation to enhance wellness. Some of the supplements that benefit cardiovascular health are vitamins D, C, B, and E; minerals such as magnesium, zinc, chromium, manganese, molybdenum, and selenium; as well as anti-inflammatory and antioxidative nutrients such as curcumin, resveratrol, N-acetyl-cysteine, and pterostilbene.

Five for Flow: Supplements for Heart Health

- CoQ10
- Magnesium
- Fish oil
- D-ribose
- Polyphenols

(See more about supplements in the Appendix, page 177.)

CHAPTER 3

Cancer

A Strong Lymph System Strengthens Your Immunity and Can Help Protect You Against Cancer Cells

Without question, the word you see at the top of this chapter is *the* word that no family wants to hear. And you don't need us to tell you how scary and complex and pervasive cancer is. That's because you've heard many of the stats: Nearly two million new cases of cancer will be diagnosed every year, and it's estimated that 40 percent of men and women will be diagnosed with cancer sometime in their lives. Cancer is the second leading cause of death in the United States.

But frankly, when it comes to this disease, it's not about the stats. It's about the people—and the fear of the unknown.

Cancer—unlike straightforward conditions that can be stopped or fixed with universal medication or procedures—comes in many forms, and there is no single way to treat it. This has been one of our medical enigmas, partly because of the varying nature of the disease, and also because cancer cells are derived from our own cells—not some outside "invader" that can be easily targeted because it is so different from our own cells.

While we can't cover cancer in all of its variations in this chapter, we can look at how our immune system works in conjunction with the lymph system to keep cancer at bay and slow the spreading of the disease. A lot of research shows that various preventive methods can help. These include not smoking, avoiding sunburn, practicing stress-managing techniques, maintaining a physically active lifestyle, and consuming micronutrients (via diet and supplementation) that have been shown to have cancer-preventing effects. There is also a genetic component to cancer, which is why regular screenings are encouraged—not as prevention, but as early detection, which increases the curative success of treatments.

Our purpose is to look at the critical role that the lymphatic system plays in preventing and slowing the spread. By understanding that, you can get a sense of how you can help your lymph help you.

By making your river of health flow strong, you have a much better chance of letting your body do what it wants to do—eliminate cancer cells before they get out of control.

Let's start by understanding a bit about the way your immune system functions.

The Inner Workings of the Immune System

To visualize how your immune system works, it's useful to think of any defense system you know, like security guards, bouncers, TSA agents, a military presence, armored knights, anything. These people have a very serious job: protect the treasure—bank vaults, airplanes, kings and queens, or anything that needs to survive unwanted advances of evil.

Your immune system has many parts that take on various roles, depending on what is happening in your body. When dealing with cancer, the treasure that's being protected is your body and your health.

Your immune system and the cells within it (though not covered in armor) are the Secret Service agents of your body. They will take biological bullets for you and die so that you can survive. Some of them also shoot biological bullets.

So let's see your immune system in action.

In the normal state of biological life, everything is wonderful. The metaphorical birds are singing, the sun is shining, and your cells are chugging and churning, keeping your organs and systems running and functioning. Everybody is happy doing their job ("hey stomach, time to digest that yogurt and blueberries!"), and your body's biological systems are zipping along 24/7 to keep you moving, thinking, working, doing, resting, sleeping, dreaming, enjoying all that life has to offer.

Sure, it would be grand if there were never any disruptions to this magical flow of life force throughout your body, but we pay a price for interacting with the world. See, when our bodies come in contact with the outside world, some of that outside world makes its way in. The outside world, as you might imagine, can enter our bodies in a number of ways—through our skin, through our airways, through our food. Some of these visitors are friendly, and our bodies welcome them.

But others? They're about as friendly as a ravenous lion, and more interested in invading our bodies and destroying things along the way.

Luckily, we have our immune system to engage in a defense process.

Our bodies identify something foreign inside us through specialized cells that act like TSA agents—checking identification. If the identification doesn't check out—meaning that the cells don't belong—your immune system kicks in. And that's when things get a little messy.

Oh, it would be nice if those pathogens—substances that don't belong in your body—would just turn around and leave. But they're not very compliant, and that's what starts a confrontation. The pathogens want to stay, and the immune cells want them to leave. You may have heard of B cells and T cells, which play various roles in the fight to defend your body. And this is where a cellular firefight begins. Your body sends white blood cells (including granulocytes and lymphocytes) to attack the area of infiltration. This causes an inflammatory response in an attempt to kill the invader.

The fallout from the fight? The area can get red and swollen, body temperature can rise, and you have an overall acute inflammatory response that keeps your body in a five-alarm state of emergency. With some pathogens, you can easily see how this fight plays out. You sneeze with allergies. You get a fever with a virus. You may spend more time hunched over a toilet bowl when the pathogens enter via Nana June's "special" sandwiches. Those responses are biological evidence of the immunity fight occurring in your body.

Typically, these run-of-the-mill fights don't last very long, because your strong and robust immune system (perhaps with the help of medication or perhaps not) can kill the invaders, and your cells go back to normal. When the firefight is finished (and some immune cells die off in the process), the "cleanup crew" of macrophages pick up the debris and transfer crucial information regarding the identity of the invaders to the "adaptive" immune system. This is what maintains immunological memory, so that the response is even quicker and more effective the next time microorganisms with those same identity papers (called antigens) show up. We call that state "immunized." All the components of your immune system then go back to their posts to be ready for the next possible invasion.

But what happens when the pathogen is more serious, more aggressive, more of an "undercover fighter," as is the case with cancer cells?

Here's where it gets tricky. Ideally, your immune system would be able to identify all cancer cells as problems. Your system does this naturally with what's called a proofreader gene that exists in virtually all cells of your body. This proofreader gene tries to identify what genes have problems (or typos) in them, so that your immune system can do its job and fight the cells.

But cancer cells, as you can imagine, are some of the most subversive threats our bodies can encounter. Because they are derived from our own stem cells, they are able to hide from the immune system, cause the blood clotting system to produce a substance called fibrin that can cover an entire tumor colony, and make our immune system think it's an innocent blood clot—and even "capture" immune cells into the tumor colony and force them to produce compounds that favor the growth and spread of this terrorist cell. Cancer cells can turn off the proofreader gene; this is like cutting power to a security camera. The immune cells might not even know these cancer cells exist—replicating, growing, and spreading. And they cut the signal that would let the immune cells come in and do their job. That leaves cancer cells alone to capture the treasure—in this case, to overrun your body with cells that kill off the normal functioning of your body's organs and systems. However, to accomplish this, tumor cells have to reach a certain critical mass and develop their own blood supply to get larger than the head of a pin. In very early stages, a healthy immune system with vibrant lymphatic flow identifies these damaged and confused cancer cells and takes them out by injecting them with deadly chemicals before they can develop into a hidden and growing threat that has cut itself off from all communication with the healthy body systems.

And this is where one of our great medical challenges lies: Cancer cells can overwhelm your immune system in the way that a terrorist group can carry out an attack by hiding its activities completely from intelligence agencies. Current therapies have looked at possible ways to deal with it—by surgically removing a tumor (which cures approximately 50 percent of cancers), killing cancer cells directly (such as with chemotherapy and radiation), or by helping to bolster the immune system (this is called immune therapy) so your body is better able to recognize and fight the menace from cancer cells that have achieved critical mass.

COVID-19 and Lymph

The coronavirus pandemic that began in 2020 has certainly made people much more attuned to the importance of the immune system, but the reality is that most people don't truly understand how nutrients—like vitamins A, C, and D—have an effect on the immune and lymphatic systems. A free-flowing lymph system helps move out unwanted substances in the body, like toxins, bacteria, and viruses. Research around COVID-19 found that those who had higher levels of vitamin D fared better than those who were deficient. Vitamin D is a key building block of the immune system.

Our immune system has two parts: innate (ready to attack foreign invaders as the first line of defense) and adaptive (works by remembering previous exposure to pathogens and attacks, as well as generates memory of a new invader). Data have shown that those the most at risk for serious health complication for COVID-19 had comorbidities. This underscores the importance of overall health. A healthy immune response is also part of what we can control when faced with new invaders that force our adaptive system to identify and fight a new problem. A healthy lymphatic system helps maintain a healthy immune response. In these times, that is more important than ever.

The cause, timing, and treatment of the symptoms of COVID-19 are actually quite different. Symptoms of loss of smell, fever, chills, sweating, diarrhea, sore throat, and cough occur a few days after the patient is infected with the SARS-CoV-2 virus. After a week or so, the respiratory symptoms become worse with shortness of breath, oxygen desaturation, and generalized inflammation.

In the first week or two, the virus is multiplying and invading the cells of the body. This varies from patient to patient depending on how disabled they are. Patients with diabetes and obesity are at higher risk because they already have an inflamed constitution. During the first phase of this disease, the best treatment is to eradicate or prevent the virus from multiplying. This can be done by hydroxychloroquine, ivermectin, and zinc or remdesivir. Sometimes there is an overlap between the two that would require both the antiviral and the steroid. After a week or two, the symptoms of respiratory distress and general debilitation are due to an overabundance of inflammatory substances or a "cytokine storm." This is not the virus but our own body overproducing substances that create lung fluid, blood clotting, generalized debilitation, and shock. Multiple studies of patients during this cytokine storm have reported both high viral counts and low viral counts and particles leading one to consider that treating the virus infection at that stage is not as effective as steroids like dexamethasone are.

However, in many cases, there is one observation that is consistent: There is a T cell depletion in this disease. T cells are lymphocytes that can produce inflammation, but specifically can also decrease inflammation and repair tissue if they are a specific T cell called a CD4 T regulatory cell. It is our belief that the T cell deficiency in this stage of the infection is of the CD4 T regs, so the inflammation increases and goes wild. Monocytes become macrophages and are attracted to the endothelium of the lymphatics and the blood vessels, causing poor transportation and communication in the lymph vessels and thrombosis and hemorrhage in the blood vessels.

The lymphatic system is integral to the transmission of immune cells, information, and substances that create a healthy inflammatory reaction and a quick reversal to an anti-inflammatory state. Patients with chronic inflammatory diseases like obesity and diabetes have severely leaky lymphatic vessels that lose immunoglobulins, triglycerides, and cells and nutrients that are necessary for a healthy immune system and intersystemic communication. Keeping those channels open and minimizing transported substances and cell loss will optimize wellness.

Lymph System and Cancer

There's a good chance that the reason you even knew anything about the lymph system before starting this book is *because* of cancer. That's the context through which you've likely most commonly heard the word—as in "the cancer was found in X number of lymph nodes that were removed with the tumor," or because you know that's one of the rivers on which cancer travels (the other being the blood circulation).

Now that you have an understanding of the components of the lymph (see chapter 1), you probably have some sense of how strong flow and healthy lymphatics play a role in cancer prevention and treatment.

As you now know, macrophages, natural killer cells, T cells, B cells, and other immune cells travel via lymphatic fluid. So if your river is stagnant, that decreases the chance that those immune cells can get to the site of where cancer cells are replicating. In the same way, the lymphatic fluid is how the renegade cells are disposed of. This is where some of the gunk is collected, filtered, and disposed of. So your immune system—to be fully operational and effective—must have its lymph running as well as possible: To send immune cells to the scene of the biological crime and to shuttle them off and trigger their self-destruction. Cancer cells develop when our stem cells become too damaged to self-repair, and certain mutations cause these now malignant cells to lose their built-in ability to self-destruct (called apoptosis), which is what normal cells do when they become too old and/or damaged. When chemotherapy and radiotherapy kill cancer cells, they do it mostly by triggering apoptosis.

Lymph is also how cancer spreads. Here's how: Cancer isn't just a single tumor that spreads by moving to another area of the body; cancer spreads when cells break

Five for Flow: Supplements for Fighting Cancer

- Vitamin D
- Curcumin
- Fish oil
- Polyphenols
- Modified citrus pectin

(See more about supplements in the Appendix, page 177.)

off from the main tumor and travel elsewhere. They're like free-running hooligans rioting in the streets of your body. Those streets are your lymph, and that's how the cells spread from the origin of the cancer to other parts of your body. It's much easier for malignant cells to survive and spread within stagnant lymph, which is not turning over new immune cells regularly. So simple health and lifestyle habits like good eating and exercise that enhance

Factoid

Research from Italy and Japan has shown that a sluggishness in lymphatics can promote tumor growth. This is likely due to a poor immune response and may lead to not only tumor growth, but the actual development of tumors.

circulation of lymph increase the chance any stray malignant cells that enter the lymphatic system will be swept away and destroyed, rather than allowed to navigate, avoid immune cells, and set up a new colony at a site distant from the primary tumor (which may be known or unknown to you and your healthcare team).

Then the problem comes when malignant cells reach a lymph node. Cancer cells in the nodes actually suppress the immune system—further increasing the cancer's chances of taking over, because the immune system is unable to successfully fight. Swedish researchers found that the cancer cells do this by disguising themselves as white blood cells. This infiltration is where new research is emerging, specifically in areas such as what attracts cancer cells into the lymph and how tumor cells suppress the immune system when they reach the lymph nodes. Most chronic problems occur at the lymph node, where this breakdown of function happens (and this is often the first outward sign of cancer—swollen lymph nodes, which indicates malfunctioning of the immune system). From the lymph node, the cancer cells can spread to other organs via the blood—and these cells, when they attack an organ, are what cause the organ to fail and are what threaten our lives.

Knowing this, researchers are exploring ways to work with the immune and lymphatic systems to understand (and thus treat) the way the nodes are invaded—and to figure out how to enable the immune system to attack cancer cells rather than be tricked by them.

What Happens If Your Lymph Nodes Are Removed?

Lymphedema is swelling that can occur in your arms and legs, commonly after lymph nodes are removed after cancer treatment. Research from the University of Singapore is showing that you can improve lymph flow through massage, proper diet that includes polyphenols and flavonoids, and even plastic surgery to remove fatty tissue that may be obstructing lymph vessels. In addition, even without nodes, lymphatic vessels will develop new channels to go around the surgical incision to the nearest lymph nodes (where fluid will be filtered).

Picture lymph nodes as railroad stations in America in 1860. Trains traveled from north and south, delivering goods, dropping off information, shuttling people back and forth, but all would arrive at a train station to carry out their function, whether it was delivery, removal, or construction. In 1860, the stations were depots for soldiers, weapons, intelligence, the return of the wounded, and prisoners. If the trains were slowed down, all these things would be slowed down. If interrupted as the lymphatic channels are in surgery, all had to be rerouted to the nearest working station, even if it meant creating new emergent railroad beds. This is analogous to the lymphatic system, where after the lymph nodes have been removed, the lymph vessels have to be reconstructed to work around the area to get to the next nearest lymph nodes. This is greatly facilitated by optimizing lymphatic drainage, as well as reducing scar tissue and fat cells that may obstruct lymphatic vessels building new channels.

CHAPTER 4

GI Disorders

An Unsettled Digestive System
Can Flow Smoothly
When Your Lymph Does, Too

Let's say we ask you this simple question: What's inside your belly?

You could answer simply: "This morning's avocado toast."

You could answer sarcastically: "Three decades of Mama's ziti."

You could answer emotionally: "A serious craving to crush half a pie right about now."

Or, as we prefer, you could answer anatomically. See, once you get just below the surface of the skin, wade through the fat, and dive through the muscle (yes, you have abdominal muscles in there!), you enter one of the most complex, interesting, and potentially tumultuous biological worlds in your body.

Your digestive system—often thought of as a series of tunnels and highways that's been deputized with the job of transporting your noodle bowl from fork to municipal sewer system—is much more than just an anatomical roadway for your meals.

Your digestive system, in fact, is an ecosystem made up of organs, organisms, chemicals, hormones, fluids, and more that influence your overall health—and how you feel day to day and hour to hour.

So you might even argue that your digestive health has as much or more importance in your overall wellness than any other area of your body. Yes, people focus on the heart, the brain, and cancer cells as the big three areas of health prevention, because they're associated with major (and scary) risks. Problems of the gut, however, often get treated like the neighbor's barking dogs: You tune them out until they *really* start bothering you.

Oh, sure, you might think about your gut in those acute times when your innards turn into a tornado of yuck—say, when you have the flu, after a night of one too many bourbons, or if you've had an unexpected encounter with undercooked poultry.

Factoid
One way to increase antioxidant enzymes and decrease markers of inflammation is through a compound called adipodren, found in natural plants like dandelion, buckwheat, butcher's broom, goldenrod, and tea. They consist of powerful polyphenols and flavonoids.

But the fact is, your gut drives your day in ways that few other parts of your body do—it influences your mood, energy, and overall wellness. And when it's disrupted, so is your life. Certainly, your digestive system is complex, and there are many possible things that can go wrong with it. Here we'll break it all down to show how your gut can become tumultuous—and how your lymph can help calm it down.

The Anatomy of the Gut

Chances are, you have a pretty solid foundation for understanding how the digestive system works. You know that food travels through your digestive system and is broken down into constituent parts that get shuttled through the body or eliminated as waste. And you probably have some idea of the functions of various parts of your gastrointestinal (GI) tract; for example, the liver is one of your toxin cleaners, and your stomach is the primary hub for secreting enzymes that help digest food. So we won't go through every step of the digestive process here, but rather highlight some of the key components and functions, as they play roles in how you might feel.

Key Organs: Some of the organs of the digestive system include the following:

Stomach: After food moves through the esophagus, it goes to the stomach, which both holds food and breaks it down (via acids and enzymes that change your bite-size taco bits into various parts that can be moved along).

Small intestine: As the broken-down food leaves the stomach, the digestion process continues in the small intestine—taking some time to move farther along and break down the food further into nutrients that are absorbed into the bloodstream. Other organs assist to produce the fluid and enzymes that help with transportation.

Colon: The trash truck of your body, the colon is responsible for getting the unused elements ready to leave your body—and where the liquid form of waste is transformed into a more solid form to pass through the remainder of your GI tract.

Pancreas: Produces enzymes that break down protein, fats, and carbohydrates, as well as produces insulin—the hormone that transports glucose throughout your body.

Liver: Cleans out toxins by scrubbing your blood and also makes bile, a key fluid to aid digestion.

Gallbladder: Plays a role by being the place where extra bile is stored, and secretes bile in response to the entry of dietary fat and/ or bitter substances into the small intestine.

Chemicals: Certainly, when it comes to the digestive system, there's a whole host of biological mad scientists making chemical stews that aid in the digestion process. These are used to break down food, to help turn it into waste, and to turn your salmon into nutrients that best feed your organs, cells, and tissues. For these purposes, there's no need to understand the function of each one of them, but it is worth noting one of the biggies: the hormone serotonin. You probably have heard of serotonin in the context of your brain—it's often referred to as one of the feel-good chemicals. Serotonin, which is transported via nerve cells, is *most* prevalent in your gut. In simple terms, that means there's a huge connection between your brain and your gut, as serotonin works to deliver messages back and forth between the brain and the gut. That, as you might imagine, makes it more clear how the two parts of the body are intertwined. Appetite, mood, energy? They have biological hubs in both the brain and the gut—with major implications about how digestive turmoil can influence your day-to-day wellness and overall health.

Vagus nerve: The cranial nerves are thus named because they directly leave the skull to service the body, rather than traveling by way of the spinal canal. They are involved in sensory perception like sight, smell, taste, hearing, and speaking, as well as receiving sensation and activating musculature. The tenth cranial nerve is called the vagus nerve because it "wanders" throughout the body (*vagus* being Latin for "wandering"). It is the longest nerve in the body and serves multiple functions.

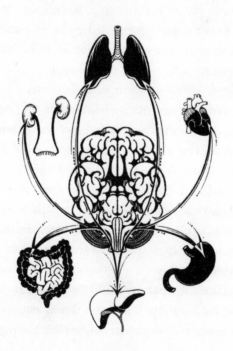

Not only is it an efferent nerve, meaning it sends signals to muscles and organs, but it is afferent as well—that is, it collects information from these areas and sends it to the brain. In fact, 80 percent of the messages distributed by the vagus nerve come from the body to the brain. The vagus nerve innervates the ear, pharynx, vocal cord, lungs, heart, liver, and abdominal organs like the stomach, small intestine, and large intestine. Furthermore, it's the nerve that supplies the parasympathetic system with innervation. The parasympathetics are the autonomic nerves that regulate GI motility, digestion, and relaxation. Serotonin balances the sympathetic system, which is the autonomic system that puts us on the alert to either defend oneself or run. The sympathetic branch of the autonomic nervous system increases blood pressure, blood coagulant ability, adrenaline, and awareness, all of which the parasympathetic system counterbalances. Sometimes with an overexpression of the parasympathetic system in an attempt to balance out the sympathetic effect, a vaso-vagal response (a drop in pulse and blood pressure) will occur, which can cause fainting or urinary incontinence.

Vagal nerve stimulation can relieve depression as well as seizures in epilepsy, suppress tinnitus, and also treat tachyarrhythmias (fast abnormal heart rhythms). Cutting the vagus nerve, a surgical procedure known as a vagotomy, was previously (more than three decades ago) performed for peptic ulcer disease. It has been observed that this procedure assisted weight loss programs (likely by reducing the efficiency of digestion) and also cut the incidence of Parkinson's disease in half in patients who had previously had peptic ulcer disease treated with surgical vagotomy. European researchers have suggested that Parkinson's disease begins in the GI tract and spreads to the brain via the vagus nerve.

Like lymphatics, the parasympathetic nerves can be stimulated by alternating cold and hot showers, relaxing environments, and food ambience. Recent research has shown that the vagus nerve actually can suppress inflammatory cytokines like IL-1, IL-6, and TNF. Doctors and researchers are eager to further explore the role of vagus nerve function in controlling inflammation through neurologic intervention, which was previously not considered possible.

Organisms: We're sure you've heard of one of the hottest words in health the last few years: "microbiome." This ecosystem of trillions of bacteria living in your gut works in a number of ways, including influencing digestion and other areas of your health. This collection of bacteria is unique for everyone, and—depending on how it functions in each individual—can play a role in any number of digestive and health issues (including immune function, inflammation, and obesity). Current thinking is that a more diverse microbiome is better for your health; this diversity is influenced by food and environmental factors, such as the air that you breathe. Antibiotics—either prescribed or as residues in animals raised for meat—as well as herbicides such as glyphosate take a major toll on the diversity and robustness of the gut microbiome.

> ## Factoid
>
> Researchers found that a high-fiber diet may improve brain health by improving the gut microbiome by producing a substance called butyrate, which can protect the brain and enhance its plasticity. The researchers also found that coffee may be helpful in Parkinson's disease, since in this disease patients had abnormal biopsies of the colon, which suggests a gut-brain connection. This may be because caffeine enhances production of the hormone dopamine, which is deficient in those with Parkinson's.

What's Wrong with Your Gut?

We say this next sentence without hyperbole: The gut is the body's greatest interface with the outside world.

Think about *that*. Your gut—how you ingest materials from outside and how the body transitions them into both helpful and harmful elements—is the main portal through which your health is affected. While there are other interfaces between internal and external (such as you skin and your lungs), the way that your gut works with those outside elements can effect *everything* that you do and feel.

Here's how to think about it: Say that you and a couple of friends want to get together for a nice evening out. You sit around the table, you laugh, you toast your friend's new job/baby/move, and you have a grand time. Everything in sync, everything works well, all systems go. Now, what happens when the dynamics of the table change? What if it's a couple of friends, a toxic boss, a friend of a friend whom nobody knows, or a stray customer who pulled up a seat and asked to share the nachos? The dynamics are *off*, nothing feels right, and it's a dysfunctional gathering.

What does that have to do with your gut (besides the havoc the nachos can wreak)?

Think of your digestive system as that table. The guests at that table include your immune system, your nervous system (remember the brain chemicals), and your endocrine system (your hormones). They all congregate in the digestive system and interact with each other—meaning that the biological dynamics can be as smooth as a gathering of close friends or it can be disruptive and contentious. And the reason it's so profound in the digestive tract is because of the magnitude of this interaction. Consider the fact that the GI system has:

- more nerve tissue than the spinal cord
- more immune cells and lymphatic tissue than the rest of the body
- more endocrine-secreting cells than any other tissue

This continuous interaction—if there is not significant coordination—can lead to calamitous dysfunction.

You can imagine that, too, right? You have three distinct systems each with a different function trying to get along with the other systems; that interplay doesn't always work (the whole idea of oil and water, the Jets and the Sharks, Congress).

In your gut, these problems don't manifest themselves in government shutdowns or *West Side Story* gang fights, but in conflicts that you feel when your innards rumble and rebel. That gastrointestinal rebellion comes in the form of inflammatory bowel disease (IBS), which is an umbrella term that covers a large array of disorders that affects the digestive tract by changing the immune system, causing inflammation, and destroying GI tissue.

The most common forms of IBS are:

Crohn's disease: This disorder occurs most often in the lower portion of the small intestine (though it can occur in other parts of the GI tract as well). You feel abdominal pain, and could also experience diarrhea (sometimes with mucus or blood), weight loss, and fatigue. The inflammation that happens here (more on this in a moment) can also erode the tissue involving the bladder, vagina, or other surrounding tissues. And it also can mean that you can't absorb some vitamins, especially the energy-giving B_{12}, which, when depleted, can lead to anemia and brain dysfunction.

Ulcerative colitis: This occurs in the colon (not the small intestine), but has similar symptoms to Crohn's—abdominal pain, diarrhea, fatigue, and the constant feeling that you have to relieve your bowels. It's diagnosed with a colonoscopy and a biopsy. A number of treatments and medications exist to control the symptoms and inflammation, as well as the use of micronutrients and macronu-trients to help correct the issue.

In both of these cases, food plays a major role in helping to avoid symptoms. You can learn what foods are associated with increased symptoms. Often, an elimination diet is used; this ap-proach has you eliminating many foods and then gradually adding them back into your diet to pinpoint which ones trigger symp-toms. Research shows that this approach can help people prevent relapses more than some of the therapies involving medication.

Now, how does this all happen? There are a variety of reasons and causes, but one that's worth noting is called leaky gut syndrome. It's a biological process that can be triggered by toxins, autoimmune and celiac disease, infections, and other things. Here's how it works:

A protein released in the digestive system called zonulin relaxes the tight connection between mucosal cells in the gut. →

That relaxation is like a tollbooth gate opening, allow-ing an avenue for harmful substances to enter, as well as fluid egress (things that go back into the gut cause GI distress). →

The immune system doesn't like what it sees, so it calls in the troops and starts an inflammatory defense to get rid of the invaders (remember, these signals are sent via the lymphatics).

When working properly, the inflammatory response calms everything down. But that's the case for acute inflammation. When this fight happens—immune system fighting an invader—acute inflammation is triggered. After the immune response destroys the invader, the body goes into a reparative phase. This is how our immune system and inflammation should work. But in cases with chronic inflammation (i.e., the firefight never calms or ends), the reparative process never happens, meaning the body can't heal.

> ## Factoid
>
> Osteopathic manipulation may also help with lymphatic tissue in the gut and lymphatic circulation. Other research showed that this technique (called lymphatic pump techniques) can help improve immune function.

Chronic inflammation is linked to the symptoms associated with the various forms of IBS. (And research has shown that this protein, called zonulin, can potentially open up the blood-brain barrier and cause inflammation in the brain, which has been linked to multiple sclerosis.)

How Lymph Works to Improve Your Gut Health

By now, you understand the importance of the lymph as it relates to inflammation: The lymphatic vessels are the channels for the ongoing cellular and molecular messages. The digestive system is even more interesting because of its unique relationship with lymphatics.

Not only do lymphatics perform the usual functions of immune support, fluid balance, and waste clearance, but they are also essential for large protein and fat molecules being transported from the gastrointestinal tract to the rest of the body. This is important because when the lymph system becomes porous and vital substances leak into other tissues, you lose the nutritional power of those substances, like vitamins, proteins, and other chemicals.

And they are also involved in the process of identifying and taking care of foreign substances that can enter our bodies through the gut.

In IBS, lymph vessels are the first cells to be presented with antigens, microbes, toxins, and other unwanted molecules. At that point, the lymphatic endothelium transmits messages to other parts of the body and then self-regulates the velocity of lymph fluid (as well as how the vessels contract and constrict). This is important because that flow determines how the lymph nodes can identify (and thus get rid of) unwanted antigens.

By that measure, the sooner these antigens can be carried by the lymphatic vessels to immune cells to establish a significant response, the sooner the inflammation can be cleared up. That's key, because if you can quiet the inflammation, you can quell the symptoms of IBS that are triggered by inflammation. Not only that, but good lymph flow provides a double dose of benefit: The sooner this is cleared, the faster that tissue can be repaired and healthy tissue can be restored. That's why the relationship between lymph and immune systems is so paramount; they need to work efficiently and competently.

In fact, Dr. Gwen Randolph of Washington University, a primary investigator of Crohn's disease, has shown that flow in the lymphatic vessels that lead to the lymph nodes are significantly reduced and obstructed in the beginning of an inflammatory bout of Crohn's disease. This creates a significant delay in the repair. Furthermore, a study from Nanjing University showed lymphatic dysfunction and an increase in endotoxins in the early stage of issues involving the blood vessels of the gut. Also, a group of researchers from Texas A&M showed that alterations in environmental factors that affect the lymphatic function can be an early factor in the development of injury.

However, several experiments in the laboratory have shown

that natural ingredients like rhubarb and peony can improve symptoms of induced colitis in mice. In a 2018 study, it was shown that the colitis was improved by a change in the microbiota (which was accomplished by a rhubarb/peony extraction).

Several studies in humans have shown that lifestyle interventions can significantly decrease toxic

---

Five for Flow: Supplements for Gut Health

- Probiotics
- Prebiotics
- Enteric-coated fish oil
- Enteric-coated peppermint oil
- Glutamine

(See more about supplements in the Appendix, page 177.)

---

inflammatory response. For example, researchers did an interesting experiment on ten healthy humans in Denmark. They injected low-dose *E. coli* toxin (without the bacteria) into the subjects and measured for inflammation response after three hours of bed rest. After several weeks, they repeated the study in the same subjects and told them to exercise as hard as they could on a bicycle for three hours. Interestingly, the level of the marker for inflammation was fourfold in the resting subjects and there was hardly a blip in response to those who exercised. This would suggest that exercise increases lymphatic flow, which helps with toxic clearance.

CHAPTER 5

Weight Management

Improving Your Lymph
Can Also Help Control Your Weight
and Fight Obesity

Even with all the weight loss products, weight loss promises, weight loss books, weight loss programs, we still have a major weight *gain* problem.

And we're not losing that anytime soon.

Weight gain—which shows in the staggering statistics of the number of people who are either overweight or obese—has become our number one health problem. That's because obesity influences *everything*.

It plays a role in heart health. In brain health. In gut health. In your overall health.

Luckily, improving lymph health can help.

Now, we could get into the variety of things that have caused or contributed to our collective weight issues, but it is a very complex and systematic problem that combines equal parts biology and psychology, social factors and economic factors, and so much more. Instead, what we'll try to do is unpack a bit about the lymphatics role when it comes to weight gain and weight loss, so you can see how your river of health can help you shed some fat if you need to (which will also have exponential benefits to the rest of the systems in your body).

55

The Anatomy of Fat Storage

The simple formula for weight gain is one you've heard many times:

> Too much food + too little exercise = you need bigger clothes

While perhaps a bit too simple, there is truth in that statement. We gain weight when we don't use up the energy that we feed our bodies through food.

However, the actual biology of metabolism and weight gain is much more nuanced. And the process reveals quite a few insights into how our bodies are supposed to work, what happens when things go wrong, and why obesity is such a health threat.

So let's take a look at how it works.

Your body—one masterful factory in totality—is actually made up of many mini-factories. Your organs. Your circulatory system. Your hormones. Your brain. So many of them.

And for you to live, you need all of those systems producing what they're intended to produce. To do those things—for your heart to beat, your intestines to churn, your brain to process these very words—you need energy. You need to fuel those factories. The way we do that is with food.

Food, in the form of calories, is what provides energy and powers all of those mini-factories.

Food—made up of the three macronutrients: protein, carbohydrates, and fat—gets broken down through the digestion process. When that energy comes in, your body releases the hormone insulin to allow energy (in the form of glucose) to be delivered into the cells throughout your body. When all the systems work well, the energy is distributed throughout, you have enough of what you need, and you function smoothly and efficiently.

But here's where it can go wrong: When you consume too much energy (hello, monster sundaes), your body cannot keep up with all the excess. So you end up becoming insulin resistant in the effort to deliver all the energy. Without that insulin to shuttle glucose, you increase the amount of circulating blood glucose, which threatens the health of your arteries, as it's one of the things that can chip away and damage arteries, triggering a domino-like process of plaque formation. In addition, any extra energy that your body can't use or otherwise burn up (via exercise, for instance) gets stored as fat.

Your body, evolutionarily, wanted to store fat so that you had an energy source in times of famine (because fat can provide energy for your body in the absence of incoming food). But since our society typically doesn't experience those periods of famine anymore, our fat never gets burned up and instead just accumulates.

As you know, extra weight is associated with increased risk of just about every health problem—heart attacks, cancer, fatigue, depression, sleep issues, inflammation, and more.

One of the reasons: Fat, especially abdominal fat, is toxic to the organs that it's near. It releases damaging and inflammatory substances that create all kinds of chaos throughout the body.

How Lymph Plays a Role in Obesity

From what you've learned so far, you could probably create a nice little picture about how lymph could help fight obesity.

Good lymph flow is all about moving toxins away to be cleared. So, yes, wouldn't it be nice if your river of health could just work like class V rapids and shuttle away a triple cheeseburger before it settles into your stomach as fat? Perhaps, but it's not quite so simple.

In a 2008 article in the *Annals of the New York Academy of Sciences*, Dr. N. L. Harvey from Australia said that despite there being observations about a link between fat and lymph, the connection has only been recently recognized. That connection: Lymph vessels—which work in conjunction with our body's absorption of lipids—share an intimate space with adipose tissue (fat). Because of that, they regulate the traffic of certain immune cells that rely on fat to fight infection. He also showed recent evidence that connects the dysfunction of lymphatics with the onset of obesity. (In addition, research from the University of Calgary shows that fat and lymphatics may be connected in the development of inflammatory bowel disease.)

Most of all, though, it's important to realize that lymphatics and fat seem to go together like sweaters and Minnesota winters. That's because all the fats absorbed from the digestive system and those created within the body itself are carried through the lymphatic channels to be processed by the liver or sent as messengers to various organs.

Three decades ago, we considered adipose tissue primarily a storage of energy and as a cushion to protect the organs and other tissue from trauma. Since then, a whole new world has opened up. We found that not only does adipose tissue have an arterial and venous supply, but also lymphatics and nerves. Through these connections, the adipose cell can receive and send messages, hormones, and nerve signals that involve all the organs, including the brain.

For example, fat cells create adipokines, or adipocytokines (from the Greek *adipo*, meaning "fat"; *cytos*, "cell"; and *kinos*, "movement"). These are cytokines (cell-signaling proteins) secreted by adipose tissue and play a role in body fat. One way: A pro-health hormone called adiponectin is a large molecule, most of which has to be carried to the rest of the body by way of the lymphatic system. If the lymph is sluggish or its channels compressed by excess adipose tissue, the hormone can be diminished in circulation.

In addition, when someone is obese (and develops the constellation of symptoms that comprise metabolic syndrome, like hypertension and high blood sugar), fats and triglycerides can leak out of the lymphatic tissues and into the fatty tissue around them. Over time, that slows the rate of lymph flow from the gut to the liver—creating serious problems in transporting vital proteins, hormones, immune cells, and other molecules that send signals in the body.

That, in turn, creates a vicious cycle of health problems where fat hurts the lymph flow and lack of lymph flow can contribute to increased fat and health problems associated with it.

How to Fight Weight Gain

Most of the tips and strategies we recommend for lymph health also have a positive effect on keeping your weight in check. And while exercise also plays a role in helping to burn energy, diet is the key driver and determinant for being able to control your weight. So please check out our chapter on nutrition starting on page 83. In the meantime, we would recommend these two approaches to start:

Concentrate on whole foods, not processed ones. The number one thing you can do to help control your weight is concentrate on whole foods—that is, foods that appear as they come from the earth. So, yes, that means fish, beans, veggies. That doesn't mean things that have an expiration date in 2043. There are two main reasons for that: Whole foods are full of micronutrients that work to improve your major systems, reduce inflammation, fight disease, and more. When eaten in the right balance (and in manageable quantities), whole food works in your favor to provide you—and all of your bodily systems—fuel. When you eat processed foods, you're not only robbing your chance of getting those vitamins, minerals, and other compounds, but you're flooding your system with toxic ingredients that cause your system to act like a schizophrenic lizard—all messed up. You'll see our full recommendations in our chapter about food, but if there's one mantra to live by, it's this: whole foods, not processed ones.

Try fasting. "Intermittent fasting" is defined in medicine as taking only water for 16 hours, coupled with eating normally for 8 hours (or other schedules such as 18 and 6, 20 and 4, 22 and 2, etc.). For example, you could simply not eat after eight p.m., then skip breakfast the next day, followed by eating a normal lunch and dinner.

Research in animals and humans has shown a benefit from such a practice. It's as if you were to stop greeting guests at your front door, so that your body has the time and energy to clean its closets. When we eat, the blood and lymph supply increase dramatically in the intestines; less goes to other organs, such as the heart. Fasting reverses this; your white cells are increasingly available to prevent and reverse injury. Try it once or twice a week to improve your risk factors for heart disease, improve your insulin response, lower your weight and belly fat, as well as lower your small particle LDL cholesterol, and your blood pressure. In addition, food tastes all the more sweet afterward—abstinence makes the heart fonder— even of the healthiest foods. Another benefit of fasting is that it increases the production of "satiety hormones" (those that make us feel full), so appetite is reduced for a while after any type of fasting, which results in lower overall calorie intake over time. Many people interpret this post-fasting effect as "my stomach shrunk," because they feel full after consuming much less food than usual.

Some people fast the whole day every week or two; others fast in an hour block over 24 hours. A 12-hour fast between supper and breakfast can be effective. An 8-hour fast is less effective, but can be recommended.

Please see the dietary recommendation in chapter 7, which will cover the holistic approach to health, weight, and lymphatic flow.

CHAPTER 6

Brain and Mind Conditions

How Lymphatics Can Play a Role in Preserving Your Cognitive Function

When we think about universal fears, a couple of things probably come to mind: snakes, public speaking, accidentally hitting "reply all." In health circles, we all have a different tolerance for being anxious about our destiny. Some live their lives with no care about the future ("Seize the day!"), while others go through routines wondering if every leaf of lettuce might have *E. coli*.

And while it's not fair to generalize about what medical issues scare people the most, many surveys show that cognitive decline— Alzheimer's and other forms of dementia—are either at or near the top of a "most feared" list.

We have seen the effects of this group of diseases, and perhaps more importantly, we *feel* the effects of them. When we observe a loved one who has lost all or much of their memory, it *hurts*. And it frightens us. For some reason, we can mentally cope with the journeys of other health fates—even cancer and heart disease. But not having one's full faculties as we age? That scares the digested broccoli out of us.

And that's not just because the stats are scary: We've seen a 90 percent increase in deaths from Alzheimer's in the first fifteen years of this century, and experts predict that the number of Americans suffering from the disease will triple by 2050 from the more than five million who have it now.

Alzheimer's and its related conditions are complicated. That's because we're really still in the early stages of learning and researching Alzheimer's. And it's also because the brain—even though it has the power to figure out how to send machines to Mars—is a universe all its own. And that means there's still more to learn about how the brain works and how it malfunctions. That said, we do know quite a bit about how cognitive disorders can happen (in a very complex way, for the record, making it difficult to treat), and we do know that the lymph can play a very powerful role in helping fortify the brain, keep it healthy, and fend off the diseases that scare us the most.

Before we see how the lymph contributes, let's drill a peephole through your skull (metaphorically!) and take a look at what's happening inside. Because when you understand how the neurological galaxy works, you'll better understand how to preserve it.

The Anatomy of the Brain

Here's a meta-statement for you: Of all the body parts we take for granted, the brain has to be the one that we don't give much thought to. After all, the brain is responsible for every verb your body performs: think, solve, decide, love, cry, move, dance, lift, speak, yell, read, remember, dream, understand the joke on page 10.

The conductor of your body, the brain sends out signals for everything to do their jobs at certain times and in certain ways and in certain circumstances. Sometimes the conductor has the whole body performing in unison, and sometimes the conductor and the rest of the band can be off rhythm. In any case, the brain is where all the anatomical magic happens—whether it's via controlling movements, or organs, or hormones, or chemicals, or any of the thousands of other things that happen in your body in any given moment.

So to try to break down all the anatomy of the brain would be a cerebrum-splitting task. That's why we won't get into everything—lobes, hemisphere, and all the different parts of the brain. But it is worth mentioning that from a big-picture standpoint the brain is basically divided into three parts:

- **Hindbrain:** Also called the lizard brain, this is the part of the brain that regulates our automatic movements and functions (think breathing, heart rate, swallowing, etc.).
- **Midbrain:** As you might guess, this part of the brain connects the hindbrain and the forebrain. But it's important for our purposes because of the role it plays in hormonal production, nerve cells, and the way that the brain communicates and interfaces with the gut (more on this in chapter 8).
- **Forebrain:** This is the largest part of the brain and includes the parts that are responsible for just about everything you can put your mind around: memory, motor function, sensory information, metabolism, hormones, sleep, and more.

For our purposes, we'll take a deeper look at the major parts of the brain that play a role in memory and cognitive decline.

The interconnectivity of neurons

Neurons: Simply, these are the cells within the nervous system, and they work by transferring information back and forth from one another. Structured with a cell body and peripheral parts (called the axon and dendrites), neurons "talk" to each other. This is how memories are formed. A message is sent from one neuron to another—the contact points between dendrites is called the synapse. You build memory through the constant chatter. We have billions of neurons in our brain, which have trillions of synapses between nerve cells, which are constantly remodeling, based on need.

You know the phrase "use it or lose it"? This happens with neurons. Let's say you learn a foreign language, you're building and storing connections between neurons to know vocabulary, syntax, accents, and the like. If you keep using the language, you will continue to strengthen those connections (and thus not "lose" your ability to order the *fromage*). But if you don't use the language consistently, those neurons will act like a couple of crazy neighbors and stop talking. Eventually, that connection will just shut down—and you'll lose what you had learned.

This communication happens in the form of electrical impulses that travel via the axon, which is covered in a myelin sheath, which helps the signal transmit smoothly. This sheath is made up of cells called glia cells. And while you may think these cells sound like nothing more than a TV show about high school music students, they're actually critically important in that they serve a variety of functions, including cleaning up debris and getting nutrients to the neurons.

It's important to know that this process of building up neurons and neural connections is a plastic one—meaning that the ecosystem of the brain is in a constant state of development and destruction. That's a natural process, as the neurons are constantly remodeling (if you're familiar with the way bones work, the process is similar—bone constantly remodels by destroying old cells and generating new ones). So it's not necessarily bad that you lose connections. You don't remember your first-grade class schedule, right? Your brain works efficiently by letting the connections you don't need any longer die off. That's just the way the neural environment works.

As you'll see in a moment, that connection between neurons and the role of glia cells is super important, not just in terms of how we form memories, but because of the role they play in Alzheimer's and cognitive decline.

Blood-brain barrier: The brain, as you might imagine, is the crown jewel of the body. And like any valuable possession, it must be guarded from hooligans who'd like to get their hands on it. For the brain, that protection comes in a couple forms. The skull, of course, is your brain's suit of armor, encasing the brain and presumably adding a level of protection against outside threats.

But what about internal threats? The ones that can creep into your brain and seek to destroy it? These are the ones like bacteria or viruses or toxins or cancer cells or any number of other threats that can chip away at your crown jewel.

That protection comes in the form of the blood-brain barrier—a semi-permeable membrane that is made up of a network of blood vessels and tightly packed endothelial cells, which act as the bridge between the outside of the brain and the inside. So it doesn't work like an open door and a closed door to have access to (or not) your brain. Think of it like a firm-yet-soft substance that provides some protection, but will let some things pass through—like a screened-in porch that keeps some stuff out (bugs) but some stuff in (the breeze).

Why does it work that way? Because your brain does need nutrients and messages delivered from the rest of your body. After they're broken down by the digestive system, nutrients need to travel through your blood vessels up into your brain to feed it and give it energy and power to run your whole anatomical operation. At the same time, it attempts to keep things like toxins and other harmful substances out. In evolutionary terms, this blood-brain barrier is perfect in theory: Let good stuff in, keep bad stuff out.

But it doesn't always work that way; that's because bad stuff

> **Factoid**
>
> In mouse studies done at the University of Texas, researchers found that melatonin assists with the clearance of amyloid plaques, a marker of Alzheimer's. This is cleared via the lymphatic system, and the melatonin seemed to have an effect on decreasing the production of toxic amyloid.

> **Factoid**
>
> Resveratrol—a polyphenol found in grapes and wine—has a protective effect when it comes to diseases of the brain, especially Alzheimer's, according to recent research. Pterostilbene—another polyphenol, similar to resveratrol, found in blueberries—is far better absorbed and three to four times better utilized than resveratrol. It supports a healthy lymphatic system by promoting healthy blood flow and circulation, nitric oxide synthesis, better control of blood pressure, and protecting the endothelial cells.

can seep in (like alcohol and drugs). And sometimes we can't get good stuff in (medication that could be helpful), which provides a challenge for today's researchers and scientists—how to penetrate the barrier when it's helpful to humans without compromising the integrity of the wall so that it continues to keep bad stuff out.

The Pathology of Cognitive Decline

In some cases of medical issues, there's a clear cause and effect: The broken bone came from the fall on the sidewalk. The black buildup decorating the lungs came from the cigarettes. The night spent hovered over the porcelain bowl came from the undercooked chicken.

With memory problems, we all can identify how they manifest themselves—we see it with short-term brain burps (what did you just say?), with forgetting routine routes while driving, with wandering off from home and getting lost.

But the cause of such issues? That's not so simple. Cognitive problems, as with many chronic degenerative diseases, have so many variables, intricacies, and layers. You can't pinpoint one bacteria, one toxin, one neuron not firing, one malfunction, one genetic abnormality, one *thing* to be the root of why a person experiences cognitive declines and why another person doesn't.

So to set up how the lymph system has a role, let's take a look at some of the biological explanations for what's happening with Alzheimer's and related diseases. As you remember from our discussion about the anatomy of the lymph system and how it plays a role in disease, three things must be in place for chronic degenerative diseases to happen. They are:

- a toxin present—something that initiates a process that insults the normal functioning of said system

- the stress by that toxin or irritant causes a hyperinflammatory response—and what is designed to help clear the toxin also makes it possible for it to attack healthy structures
- poor clearance of the toxin, meaning that the toxin stays in contact with the tissue long enough that the inflammatory response increases—and thus causes more potential harm because it doesn't shut off or clean up

That also holds true for brain-related conditions, as all those things can be taking place. So let's look at some of the main ways that the brain can malfunction and then trigger a series of reactions that lead to the mind-jumbling diseases of the brain.

Also, to be clear, it's important to recognize that there is a genetic component, as it relates to a gene variant called ApoE4, which, if present, increases your risk of Alzheimer's significantly. This gene came into being in the hunter period of human development (when people were more likely to be wounded or have an infection); it was developed for inflammatory protection. The average life-span some fifty thousand years ago was thirty years; the body didn't care about any long-term damage of inflammatory response. But as we've increased our longevity, what was designed to protect us can backfire—as the inflammatory response over time has become harmful to our bodies. If a person has a gene with two ApoE4 alleles (one by the mother and one by the father), their chances of developing Alzheimer's is about 50 percent. We all have two alleles, but not all have the ApoE4. Your risk drops if only one allele has it. The reason: The allele is associated with an increased number of plaques called beta-amyloid that can build up in the brain.

Beta-amyloid is part of a protein called amyloid that your body naturally produces. This protein fragment has a purpose: to help and support neurons and those connections between neurons (the ones that are responsible for helping you remember information).

In older adults, those beta-amyloid fragments can develop into plaques that slow and stop the connections between neurons. In addition, they can cause what are called neurofibrillary tangles—where the connectors of each neuron mix with one another (remember those dendrites, which are the little branches off the cell that connect with each other).

So what happens when the beta-amyloid either clogs, cuts off, or tangles the connections between neurons? That's right. You can't remember that we just used the word "dendrite." Over time, the amyloid effects the tau proteins, which are fine protein extensions of the dendritic cell; as this progresses, the dendritic cells die.

Think about all the wires going from your wall to your TV system or power lines along the roads. Those lines are meant to communicate from one source to another—and you need clean lines in order to send and receive messages. Well, if someone came in and cut your power cord to your TV, you wouldn't be able to watch *House of Cards*, or if a tree falls on a power line, your electricity would go out. That's essentially the same thing that's happening in your brain: Something is causing the connections to be lost or to malfunction, so the neurons can't communicate—and the person can't remember x, y, or z.

Now, one of the hallmarks of Alzheimer's is that patients show increased signs of inflammatory markers—meaning there's an inflammatory response happening in the brain—as a way to handle insults to the brain. When the brain is insulted against some kind of toxin (environmental, biological, nutritional, bacterial), it creates an inflammatory state, where cells from outside the brain enter the blood-brain barrier, which then compounds the inflammation in the area.

One prime example is with the excitotoxin called monosodium

glutamate (MSG), which is found in a variety of foods. MSG can pass through the blood-brain barrier and damage brain cells and even be a trigger for cancer (much of this work was pioneered by neurosurgeon Russell Blaylock). MSG is widely used in foods because it stimulates the brain to think it is tasting something good; you probably know it as a common additive to Chinese food, but it's also widely used by the processed food industry—many processed foods simply would not taste good if the glutamate was removed. When there's

> **Factoid**
>
> Broccoli and cauliflower may help prevent Alzheimer's disease. Recent research shows that a substance they contain called sulforaphane can enhance brain-derived neurotrophic factor (BDNF), which supports the health and growth of neurons and synapses (where messages travel between neurons).

too much glutamate, neurons fire over and over—and that overuse leads to premature nerve death. This is called excitotoxicity. The brain has trouble keeping up with the cleanup, and thus increases the risk of damage. When all this damage occurs and police cells can't patrol and clean up, this aggravates the area and furthers the inflammatory response. That also underscores the importance of lymph channels to help clear the brain of toxins, of damage, and of plaques that build up in the brain.

So remember the blood-brain barrier? The toxins are coming in via that semipermeable entrance—and staying there (for some toxins, it could be easier to get into the brain than to get out). And the chronic exposure to the toxin is what triggers the inflammatory response and thus possible attacks on healthy-functioning parts of the brain.

So where does the lymph system come in? We'll discuss this in detail in a moment, but the short answer is this: What do you think helps clear the beta-amyloid plaque or toxins from the brain? Yes, your secret river of health is your biological cleanup crew that works to free the debris, which can help untangle the wiring and keep connections intact. High-functioning lymph helps remove the toxins, inflammatory signals, and beta-amyloid—which are the reason that those with cognitive decline literally get their wires crossed.

The Lymph and Brain Issues

You've already seen the lymphatics in action with other parts of the body, and the role here—the path through which the body delivers info and clears toxins—is similar, but specific to helping preserve cognitive function.

The most significant cognitive disorders we face today are Alzheimer's disease and Parkinson's disease. Both of these diseases (and perhaps multiple sclerosis and the autism spectrum as well) involve inflammation caused by the poor clearance of metabolic toxins.

While the cause of both Parkinson's and Alzheimer's is not clear, they do have some characteristics in common. One main one: Both show excessive production of substances called beta-amyloid, tau proteins, and alpha-synuclein. Here's what they do:

- Beta-amyloid is the predominant abnormal peptide found as plaques around neurons in Alzheimer's. While it is possible to have beta-amyloid and not have Alzheimer's (meaning that you can't diagnose someone with Alzheimer's just because they have this plaque), the glymphatic system is what's responsible for clearing this waste from the brain. The big problem happens when they're not cleared and they clump together abnormally.
- Tau protein exists within the cell, and it's responsible

for holding together the microtubules, which carry substances within the cell from one area to the other. But when they detach from the microtubule, it disintegrates. One theory for the onset of Alzheimer's is that the death of the neuron is caused by the separation of the tau protein from the microtubules.

- Alpha-synuclein is a protein normally found at nerve synapses to help with the transmission of messages between neurons (which is what makes up your brain function). However, this protein can become dysfunctional, multiply, and cause cell death, which may lead to Parkinson's and Lewy body dementia.

Several thoughts about the possibility of the origins of Parkinson's disease exist, but it's clear that lymphatic clearance is essential to remove toxins as early as possible and to keep the patient well hydrated so they have a well hydrated cerebrospinal fluid system (some research shows that Parkinson's is associated with low cerebrospinal fluid volume).

Another theory on the cause of Parkinson's disease involves the possibility of gut viruses creating an abnormal relationship with gut bacteria, which then leads to a lower production of dopamine (because the bacteria that are removed by gut viruses are the ones that help create dopamine). Another study revolved around the role of the leaky gut. As certain proteins produced by gut bacteria permeate through the leaky gut and also leak through the blood-brain barrier, it's thought that

Factoid
University of Alabama researchers looked at brains of mice and found that most of the bacteria in them were ones commonly found in the gut. They theorized that the bacteria indeed didn't originate in the brain, but traveled from the gut to the brain (thus underscoring the connection between the gut and the brain).

this can trigger the production of beta-amyloid (and doesn't stop until the leaky gut is healed).

My good friend and colleague Dr. Russell Blaylock is also working with us on another theory. Besides being the author of the monthly integrated medicine newsletter *The Blaylock Wellness Report*, Dr. Blaylock is the world-renowned neurosurgeon who coined the term "immunoexcitotoxicity," describing the involvement of the microglia in a ready state of primed hyperreactivity that potentiates the inflammatory and excitatory response initiated by the immune system in the brain.

An inflammatory response in the neuroimmune cells is effected when a beta-amyloid molecule combines with a toxic amino acid, creating a highly inflammatory substance. This toxin is cleared only through the lymphatic system and is potentially lethal to the lymph channels as well as to neurons. We propose that if lymph flow is not robust or is delayed, prolonged contact can lead to lymphatic dysfunction and poor toxin clearance and neural cell death.

These are some of the reasons why the inflammatory/excitotoxic response is so important. In your brain, glia cells are responsible for initiating the inflammatory response to fight toxins or trauma. But sometimes, immune cells from outside the brain—such as B and T cells (and especially monocyte/macrophages)—can enter the brain and trigger inflammation. Once the macrophages enter the brain, they take on the full appearance and function of microglia. These cells, as well as the toxins, can pass through a leaky blood-brain barrier and can also enter via the part of the brain having little or no barrier—called the circumventricular organs of the brain.

Researchers have recently found that there's a way to flush the waste out of the brain; it is called the glymphatic system because it involves those glia cells and lymphatic function. Much of the credit for this discovery goes to Dr. Jonathan Kipnis of the University of Virginia, as well as Dr. Maiken Nedergaard, who has studied the role of lymph in the brain.

For many years, it was believed that the central nervous system did not contain the lymphatic system, but Dr. Kipnis and his team found, in mice, that vessels contained markers for lymphatics— and upon further examination, found that the lymphatic tissue connected with the glymphatic system as one of the ways to remove toxins from the brain. This drainage works through the nasal mucosa and lymphatics of the cranial nerves.

The lymph works to drain the waste. It takes beta-amyloid out via cerebral spinal fluid to large veins—a process that is most active during the rapid eye movement (REM) cycle of sleep. It takes place for 15- to 30-minute cycles per night, and during this brief time period, about 70 percent of the clearance of amyloid plaques takes place. Research suggests that better clearance is associated with sleeping on your side rather than on your back.

However, this system can't always work well—sometimes because there's not enough lymphatic circulation (which is why it's important to practice some of the methods we describe for improving lymphatic flow) or there's too much of a buildup of toxins. Researchers have found that buildups of debris associated with Alzheimer's have been shown in cerebral-spinal fluid in patients with the disease, indicating the importance of clearance of this debris. The lymph would help clear the debris, take it through the lymph nodes, and then in lymph paths to the liver so it can be metabolized and eliminated. This early research suggests that lymph is a vital component for getting the brain clean and clear of the substances that can gunk up neural connections. This cleansing process is important not only in Alzheimer's disease, but also in other neurologic disorders like Parkinson's disease (where synuclein is the offending protein) and in multiple sclerosis (where it's been shown in mice that symptoms of MS can be exhibited by manipulating lymphatic drainage).

So where are the tension points then? Well, if you don't have enough cerebral spinal fluid, you may not have enough to wash out the toxins. This is why it's important to drink plenty of pure, clean water. About 17 ounces of spinal fluid is created a day, with about 7 ounces floating around in the brain cavity. In addition, a tight blood-brain barrier may keep toxins from entering the brain, but it could also close the gate, so to speak, to prevent some of the toxin buildup from an inflammatory response by the glial cells of the brain from getting back out. Important note: The cerebral spinal fluid does not travel via the vascular system through the blood-brain barrier to the brain; there are separate channels for this fluid to travel into the brain tissue.

The main point here is that we need two-way flow—where the lymph can help deliver information about toxins (to help the immune system do its job) and then clear those inflammatory cells. We don't want them hanging around for too long, because that's what makes the environment unhealthy (because immune cells can start attacking healthy cells if they stick around). This is the role of lymph—being the river by which this two-way flow must happen. Get helpers in, and take potential harmers out.

Improved lymphatic flow—through exercise, diet, sleep, and hydration—increases the chances of cleansing brain tissue and lowering the risk of developing brain-related conditions related to toxins and the inflammatory response within the brain.

Factoid

Diet is important not just for fighting fat. In a Harvard study that looks at diet and the mental ability of 28,000 men over the age of twenty, those who ate six or more servings a day of vegetables had better memory. Those who ate at those levels had a 34 percent lower chance of developing memory issues compared to those with the worst diet.

Brain Building Via the Lymph

Later in the book, we'll explore all the ways that you can improve your lymph in detail, and we've already mentioned them in the context of other diseases. These methods seem to have an effect on the lymph and can be especially beneficial for the brain. It's also worth noting that some interesting recent research is looking at lymphatic drainage as a way to treat neurologic diseases—by maximizing its ability to clear waste and help maintain water balance in the brain—with perhaps medicinal approaches to improve lymphatic flow. But in the meantime, these are the best approaches for helping your lymph help you:

Exercise. The jumping action of a mini-trampoline is especially helpful because of the way lymph travels. Jumping allows you to push the fluid up—and the one-way valves stop it from immediately traveling back down via gravity. When you get moving up and down, your lymph gets moving up and up.

Flavonoids and polyphenols. These substances—found in virtually all fruits and vegetables—have been shown to have protective effects throughout the body. They've also been shown to have influence on the brain. For example, some research has shown that flavonoids may significantly decrease oxidative stress in the brain. And a recent study in rats has shown that flavonoids can help reduce beta-amyloid, and we know that flavonoids do help increase lymphatic flow and function, mean-

---

Five for Flow: Supplements for Lymphatic Drainage

- Magnesium theorate
- Phos choline
- Polyphenols
- Phos serine
- Methylated B complex

(See more about supplements in the Appendix, page 177.)

---

ing that fruits and veggies are an important part of your anti-Alzheimer's arsenal.

Skip the Grains. Gluten—which has been shown to cause problems in the gut for some people—is also suspected to have toxic effects on the brain as well. Our diet recommendation is to stay away from gluten and grains, and to emphasize fish, vegetables, fruits, and oils. In our recipes, you'll see many options for gluten- and grain-free meals. Gluten is the most inflammatory of a group of proteins contained in all grains except millet and sorghum, known as lectins. If lectin proteins are absorbed through a leaky gut (which normally would prevent their absorption), they cause inflammatory reactions throughout the body, including the brain. If the gut is leaky, the blood-brain barrier tends to be leaky as well.

Protect your microbiome. There is some evidence to support the theory that certain neurologic diseases are caused by an overgrowth of pathogenic bacteria in the gut. Antibiotics and refined foods destroy healthy dopamine-producing bacteria that are replaced by pathogens, thus leading to neurologic issues.

Sleep on your side. As discussed, this seems to be the best position for aiding with lymphatic flow in the brain to help clear it of the gunk that's developed as part of the inflammatory process.

Increase your BDNF. Until recently, it was thought that no new brain cells could be grown. That changed with the

> ## Factoid
>
> In a study in Australia, it was found that those with significant deficiencies in vitamin D had a higher risk of schizophrenia. Those with low D were 44 percent more likely to develop schizophrenia (this is an important point for pregnant women: to eat foods with high levels of D, enjoy sunlight, and consider supplementation).

discovery of a protein called BDNF (brain-derived neurotrophic factor). When activated through epigenetics, neurons grow, as do connections between the nerve cells to repair damaged cells. Many things can stimulate the expression of BDNF, including:

- Aerobic exercise, weight lifting, and regular walking.
- Quality time in the sun.
- A traditional Mediterranean diet and foods high in polyphenols and flavonoids, including blueberries, green tea, olive oil, black pepper, turmeric, chocolate, and omega-3 fatty acids.
- Prebiotics (insoluble foods that good bacteria feed on), such as garlic, lentils, mustard greens, onions, tomatoes, bananas, asparagus, barley, and leeks.
- Chewing your food well (chewing is a BDNF stimulator!).
- Herbal remedies such as ginseng, ashwagandha, cordyceps, rhodiola rosea, and bacopa.
- Butyrates/butyric acid, a short-chain fatty acid that is metabolized from insoluble fiber in the colon by good bacteria and heals the colon; also stimulates nerve regrowth.
- Compounds like magnesium, quercetin, resveratrol, niacin, zinc, and N-acetyl cysteine.

What to Eat, What to Avoid

Eat These	Avoid These
Plant-based foods	Trans fats
Crucifers such as broccoli, kale, cauliflower	Sugar
High-fiber foods like flaxseed	White flour
Turmeric, cayenne, garlic	Artificial coloring
Mushrooms	Dairy
Berries	Processed meats
Green tea	Excess fat
Pure water	Omega-6 in meats
Fermented soy	
Olive oil	

You Can Grow New Neurons

It appears that we may be able to grow new neurons through a protein called BDNF (brain-derived neurotrophic factor). It seems to work by messaging genes in nerve cells to regrow, repair, and reconnect them. How do you boost your BDNF? Some things that have been shown to have such an effect: exercise, stress management, eating unprocessed foods (high sugar and saturated fat decrease BDNF), and emphasizing flavonoids (like berries, chocolate, tea, olive oil, and black pepper). Some supplements are also worth considering, such as ginseng, curcumin, quercetin, butyrate, and niacin/niacinamide. Sun exposure also seems to be helpful.

CHAPTER 7

Fuel for Flow

The Optimum Diet for Lymphatic Health

Throughout history, we've survived all kinds of polarizing conflicts—north vs. south, left vs. right, Red Sox vs. Yankees, Rolling Stones vs. the Beatles. Many of our debates, of course, are simply a matter of taste and opinion. But when it comes to science, things can get a little murkier, yet stay just as polarizing.

Diets fall soundly into this category. These days, there are so many conflicting reports, pieces of advice, and full-fledged industries built upon diet and nutrition that it can be hard to sift through the mountain of information—and know what's really right, what's really true, and ultimately, what you should do.

There are some certainties: For one, it's clear that too many treats will probably require you to have plenty of treat*ments*.

But there are also many variables: For example, not every approach works the same in every single body. Plus, while there is a lot that we do know about how food interacts with the body, there's also plenty that's still as murky as a muddy pond. While that makes things difficult to manage, it also means that you have to learn as much as you can—to sift through the argumentative culture of dieting to figure out what is going to work best for you and why. This is especially difficult, of course, in an environment when tough-on-your-body foods are readily available, relentlessly tempting, and easy to grab.

So what we're going to do in this chapter is to break down the science of nutrition—and how to directly influence your secret river of health. We'll look at macronutrients, micronutrients, how certain diets can interact with your lymphatics.

Our main goal here is to equip you with the most potent approach to improving lymphatic flow, so that you can achieve all of the benefits—and do so simply with the medicinal power of food.

The strategy is two-fold: 1) Avoid foods that clog or inflame your vessels, the ones that carry your blood and the ones that carry your lymph. Most "comfort food" should actually be labeled "discomfort food" because of its disease-causing potential. 2) Choose the rainbow-colored foods that will leave you light and refreshed, whistling with energy and joy and keeping your lymph (and the rest of your body's systems) running strong. Best of all, you can do this *and* have fun with food, so that your tongue is as satisfied as the rest of your body.

Macronutrients, Micronutrients, and the Standard American Diet

Certainly, there's enough science about nutrition to fill many textbooks, so our goal here isn't to inundate you with biochemistry and a million variables that affect how food works inside the body. And there's a good chance that you already know a bit about nutritional sciences. Before we dive into the guts about how various diets influence the gut, it's worth reviewing some basics. The four main areas you should know about:

Caloric Energy: Every ingredient is used to fuel our body. This energy comes in the form of calories. What we use is burned off; what we don't is stored in our body or expelled. However, every calorie is not created equally. That's because of something called macronutrients.

Macronutrients and Micronutrients: Macros come in three forms: protein, fat, and carbohydrates. And they all have different functions. Protein, for example, serves as the building blocks for our cells, while carbs and fat are used as energy (albeit in different ways). Micronutrients are vitamins, minerals, and compounds found in food that also have effects on the body, namely in terms of promoting good health and strong systems at the cellular level.

Food's Effect on the Body: While it's not quite as simple as saying some foods are good and some aren't, it is reasonable to say that most foods do skew one way or the other when it comes to how they interact with the body. For example, simple sugars (especially when consumed in large amounts) tend to wreak havoc on multiple systems, including increasing damage to blood vessels and promoting damaging (chronic) inflammation throughout the body. Other foods (both macro and micro) have a positive effect, calming down inflammation or clearing blood vessels, for example.

Whole vs Processed: In general, the more you can eat food the way it came from the earth, the better. And the more processed foods you have, the worse effects they can have on your body.

So, you're probably familiar with the standard American diet (indeed, it is sad!). You can even picture the plate: burger, fries, and shake or any one of its derivatives (even global adaptations like spaghetti and meatballs or #TacoTuesday). Bottom line: It's filled with saturated fat (the bad form of dietary fat) as well as added sugar—and the many extra calories that are associated with them.

And you probably already know that obesity—one result of diets high in saturated fat, sugar, and calories—is associated with heart disease, diabetes, and other chronic disorders. This can also be seen in epidemiological studies in Japan, the Hawaiian Islands, and California, where the rate of arteriosclerosis, for example, changed in ethnic Japanese who migrated from Japan to California. The same was true when the Irish moved to the United States and Greenlanders emigrated to Denmark. Furthermore, during World War II, in a period of extreme stress, Danish citizens were forced to eat grains and vegetables rather than their usual high meat and dairy diet. The incidence of heart disease dramatically dropped during this period, only to increase when the war ended and they returned to their original diet.

The standard American diet, frankly, is a killer diet, directly related to degenerative disease such as heart disease, cancer, diabetes, and other illnesses that have become so common they are considered to be simply part of growing older. However, these and many other degenerative illnesses are largely consequences of what we eat.

Why has this happened? Four major reasons:

We Eat the Wrong Kinds of Fats: A typical American diet averages 34 to 40 percent fat. Even as far back as 1988, the Surgeon General's Report on Nutrition and Health attributed 35 percent of cancer deaths to diet, saying dietary fat was linked to them. The American diet is full of the so-called "bad" fats, saturated fats derived from animal and dairy sources and fried foods (oxidized, also known as "rancid" fats). Studies have shown that even fats once thought to be healthy, such as margarines and other trans fats, actually promote degenerative disease.

The American Heart Association and National Cancer Institute recommend that the diet include 30 percent of its calories from fat. However, a growing body of research indicates that this level may be too high to prevent heart disease and cancer. It appears that a diet approximately 15 to 25 percent of calories from fat is optimal for prevention, provided these fats are high in omega-3 fatty acids and low in omega-6 fatty acids, and free of trans fat. The ratio between omega-6s and omega-3s should be about 3–6 to 1. Our current diets are more like 25 to 1.

We Don't Get Enough Fiber: We know that the word "fiber" is something you may associate with old age, constipation, or both. But the fact is that Americans don't get enough of this colon-cleansing form of carbohydrates. This low-fiber diet is directly related to our high incidence of hemorrhoids, varicose veins, diverticular disease, and cancer of the colon. Furthermore, the American diet, based on meat and dairy products (neither of which have fiber), contains more than twice the amount of protein considered to be health supporting. The milk protein casein has been implicated in cancer of the prostate and colon as referenced by Colin Campbell, author of *The China Study*. Excess protein can cause calcium to be lost in the urine, and this may be associated with kidney stones. The liver and kidneys are also taxed by protein overload. Four ounces (120 grams) of protein-rich foods daily is sufficient for a healthy adult. Most Americans eat this amount three times a day. Instead, we should be swapping in vegetables, beans, fruits (which have fiber)—and cutting back on some of those protein-packed foods (like meat and cheese). One note: Protein is not considered to have a negative effect on the body; it's that we tend to consume too much of it, often at the expense of other macros.

We Consume Too Much Sugar: High sugar isn't just associated with colorful cereal boxes. Sugar is found everywhere—in processed foods, baked goods, sauces, and more. And we simply eat too much of it. Too much sugar is believed to contribute to coronary disease, diabetes, cancer, and infections. High sugar consumption results in high insulin levels, which leads to weight gain, and further insulin resistance, in a vicious cycle, which often culminates in diabetes, coronary disease, and dementia. Add this to all the processed and fast foods laden with additives, colorings, preservatives, hormones, pesticides, and antibiotic residues, and it is no wonder that Americans experience ill health on a large scale.

We Don't Get Enough Micronutrients: The micronutrients you need include vitamins, fatty acids, minerals, and amino acids. These elements may be low in your system due to past poor dietary choices, as well as the farming practices of not replacing key elements to the soil, except in organic farming. For example, the conventionally grown apple today contains 60 percent less magnesium than in the 1930s. It's estimated that 70 to 80 percent of Americans are low in magnesium.

Deficiency of micronutrients of any kind can cause serious illnesses, which is why we recommend supplementing your diet (fruits and vegetables are the main source of key micronutrients). You will see a complete list of our recommendations and dosage guidelines in our plan starting on page 177, though you will want to discuss this with your doctor.

Since evidence shows that the standard American diet is associated with coronary artery disease (CAD), attempts have been made to limit cholesterol and saturated fat intake. However, moderate reduction in fat has proved disappointing for overall CAD prevention. Some diets recommend an even greater reduction to only 10 percent. Americans have substituted some polyunsaturated fat for saturated fat, but the percentage of protein has remained constant (except in high-protein diets), and sugar consumption has gone up dramatically, to the remarkable statistic of 152 pounds of sugar sweeteners per person in the United States each year.

The bottom line: Even as the evidence has pointed us in the right direction, so many of us are still going in the wrong one.

The Best Diet for Lymph Flow

For your lymph system (and your overall health and wellness), we, as well as many nutritionally focused physicians, recommend the Mediterranean Diet. By adopting this diet you can increase the strength of your lymph flow as well as your heart health, reduce your risk of cancer and diabetes, lower your chance of stroke, decrease your blood pressure, maintain an optimal weight, and get dramatic relief from many other common ailments.

Meats and sweets:
less often

Wine: in moderation

Poulty, eggs, cheese, and
yogurt: moderate portions,
daily to weekly

Fish and seafood: often, at least
two times per week

Drink Water

Fruits, vegetetables, grains (mostly whole), beans, nuts,
legumes, seeds, herbs and spices): base every meal
on these

Be physically active

Whole Fresh Fruits	Melon, citrus, berries, grapes, apples, pineapples, and bananas. Fruits, rich in fiber, provide many important vitamins and minerals, including Vitamin C and potassium.
Whole Fresh Vegetables	Dark-green leafy vegetables such as kale, collards, turnip green, and bok choy. Cruciferous vegetables (the cancer fighters) include broccoli, cabbage, cauliflower, Brussels sprouts, mushrooms, sweet potatoes, squash, carrots, onions, peppers, and garlic. Various vegetables are stars in the current firmament of research for their anticancer properties, their generous amounts of fiber, and remarkable amounts of vitamins and minerals, such as vitamin C and beta-carotene.

Whole Fresh Grains	Whole grain bread, tortillas made with whole grains, brown rice, whole grain pasta, hot and cold whole grain cereals such as oatmeal, corn, millet, bulgur wheat. Whole grains provide a rich source of fiber, vegetable protein, and vitamins and minerals, especially zinc and B vitamins.
Monounsaturated Fats	High-polyphenol olive oil; also avocado and coconut oils. Avocados and nuts.
Mostly Plant-Based Proteins and Omega-3-Rich Protein Sources	Dried beans, peas, and lentils, avocado, walnuts, flaxseeds, and legumes are a fine source of vegetable protein, vitamins and minerals, especially calcium, zinc, and the B vitamins. Soy has now been deemed anticancer by five studies that showed that breast cancer patients fed soy had less recurrence of their disease.
Fish and Seafood	Wild salmon, tuna, mackerel, haddock, shrimp, mussels, sardines, anchovies, clams, oysters, and scallops.
Poultry, Eggs, Cheese, and Yogurt	Moderate amounts of pastured chicken, free-range eggs, goat and sheep cheese, and yogurt.
Meats and Sweets	Sparingly.

In our plan, we'll take you through the specific ways that you can use the food groups above to create lymph flow and taste-friendly meals that keep you satisfied, happy, and healthy.

Food Facts: Why Nutrition Isn't Always Easy— and What to Keep in Mind

As we said at the start of this chapter, nutritional information is rarely one size fits all. So it can be difficult to not only navigate the unclarity, but it can also be misleading. We recognize that the easy-to-follow charts above and below are a good guiding post, but they're not meant to be *law*, because individual circumstances

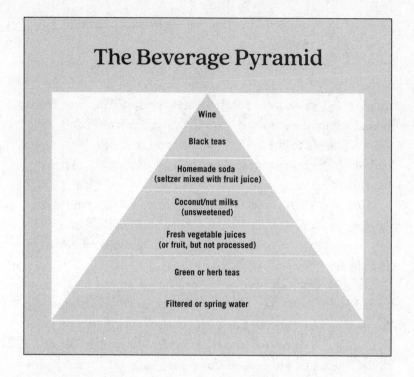

The Beverage Pyramid

Wine

Black teas

Homemade soda
(seltzer mixed with fruit juice)

Coconut/nut milks
(unsweetened)

Fresh vegetable juices
(or fruit, but not processed)

Green or herb teas

Filtered or spring water

dictate individual decisions. That said, we are comfortable saying that our lymphatics diet above is a scientifically sound approach that has the potential to pay enormous health dividends.

However, we want to be mindful of nuance and customization, so we are including a variety of approaches that may play a role in how you think about food.

"High-Carb" Can Be Confusing: Say the phrase "high-carb" and for some people it sounds as evil as "puppy strangler." But that's because people often confuse and misunderstand simple versus complex carbohydrates. Examples of simple carbohydrates include starches like potatoes cooked in the form of French fries or potato chips, fast and processed foods, and added sugars. These can cause

serious health problems by elevating triglycerides, fat, and choles-terol levels in the blood.

Complex carbohydrates, however, are found in foods such as beans, vegetables, and grains and most often do not raise the glucose and subsequent insulin level. The sugar in these foods, bound within the cellulose, takes time and digestive energy to be-come available to the body. The body can thus process complex carbohydrates more appropriately because of the time and energy expended.

However, even some vegetables such as carrots have a high "glycemic index"; that is, the glucose becomes relatively more rap-idly available to the body and can sometimes cause problems, es-pecially in people with diabetes.

Sucrose, fructose, and lactose are simple sugars rapidly con-verted by the body into glucose, the immediate source of energy for metabolism. Glucose that can't be used immediately or stored as glycogen is converted into triglycerides, the storage fats of the body. These fats are broken down over time into small hydro-carbon fragments used in mitochondrial respiration for delayed energy production. Triglycerides are the most abundant intracel-lular and extracellular fat in the body. Elevated triglycerides serve as a known risk factor for heart disease, elevated LDL, and obe-sity, and the culprit behind elevated triglycerides is often over-consumption of simple sugars.

Consider the Mediterranean Way: If you already eat the Mediter-ranean style (fish, vegetables, olive oil), there's evidence that it's a good one to stick with, given the emphasis on fruits and veg-etables, as well as the healthy fats in fish and olive oil to assist with lymph drainage. Some folks, especially those with neurologic dis-orders, have adapted Mediterranean with keto (higher in fat, lower in total carbohydrates) and have had some success.

Now, what makes things tricky is that if you have other issues—like cardiovascular disease—it is advised to follow a mostly plant-based, moderate-fat approach. This is why you have to use some judgment, and work with your doctors and healthcare team to adjust the diet that best suits your needs.

Some No-No Foods May Be A-okay in Moderation: For example, ingestion of small amounts of wine and coffee seem to be protective against the development of Parkinson's disease. And cocoa

Principles and Components of the Mediterranean Diet

- Enjoy meals with others
- Maintain an active lifestyle
- Drink plenty of water
- Incorporate brightly colored fruits and vegetables with each meal
- Include olive oil as your principal source of fat, along with avocados and nuts (avocado and coconut oils are also beneficial).
- Enjoy fish and seafood at least twice a week
- Eat poultry, eggs, and goat and sheep cheeses in moderation
- If desired, red wine with a meal may be enjoyed
- Consume red meat, saturated fats, and sweets less often than other foods

and dark chocolate polyphenols change the microbiome of the gut and can have an anti-inflammatory effect.

Substitutions Will Save You: When you are confronted daily with a decision about what to chow down on, seek substitutes that will not clog or inflame you. Instead of the donut (appropriately labeled do-not), go for the crunchy apple with a few walnuts, and top it off with the lymph-detox smoothie given in the Appendix. Instead of the devil's food cake (again, rightly named), go for some raspberries mixed with pomegranate seeds, and the radish jicama topped with blueberries. Make your dietary choices fun, or you won't stick with them. Take it as an artistic and creative challenge: pair them with exercise and yoga, and they will make the rivers of life within you sparkle with the joy of lifelong health.

You Usually Won't Go Wrong with More Veggies and Fruits: These are nature's most powerful medicine, as they are filled with polyphenols, flavonoids, and fiber. And they're rich in micronutrients that can increase their lymphatic flow. We recommend more vegetables than fruits. Some fruits are high in sugar; low-sugar fruits such as apples, berries, and avocados are preferred.

What Kind of Eater Are You?

More and more Americans are becoming aware that meat loaf and mashed potatoes aren't the best dietary pathway to heathy living. Many know that the standard American diet is about as good for you as driving drunk without a seatbelt. That's why many people have turned to other dietary approaches to prevent and reverse disease. Some of these popular diets are:

94

Vegetarian: This diet consists of vegetables and fruits and omits any meats or meat products. A vegan diet allows for no animal products, and an ovo or lacto vegetarian diet allows for eggs and milk or cheese.

The vegetarian diet has been studied most extensively and for a longer period of time than any other diet, so choosing a vegetarian diet is a safe bet for health. The only caveat is that B_{12}, which is obtained from meat, has to be supplemented.

Lectin-free: This diet avoids most grains, uses only pressure-cooked beans, and requires peeling and seeding of vegetables that have seeds in them, such as tomatoes and peppers. Lectins are defensive toxic molecules that plants have developed over millions of years that are poisonous to animals, which try to eat the fruit, leaf, or stem of the plant. Gluten (gliadin) is a lectin protein found in many grains—wheat, rye, oats, etc., that causes celiac disease by opening the connections between the cells of the GI tract and allowing toxins to come in and fluids to egress. This is a very restrictive diet—no grains, seed-bearing plants such as tomatoes and eggplant—and somewhat difficult to comply with.

Mediterranean: This diet, which was typified by the inhabitants of Mediterranean countries like Turkey, Greece, Italy, and Spain, was more properly organized as the specific Mediterranean diet in the early 1990s by Harvard's Dr. Walter Willett. Along with various fruits and vegetables, certain types of fats, seafood, whole grains, yogurt, hummus, tabbouleh, and nuts and seeds, the diet features many spices and herbs that have benefits to the lymph system. These include turmeric, ginger, and astragalus.

A very significant feature of the Mediterranean diet is that there is very little meat—particularly less red meat—than in the standard American diet, and few processed foods, especially

How to Supplement

Supplementation of micronutrients is worth considering to make sure you get all of the health (and lymph) benefits they provide. Our recommended list includes:

- Omega-3 oils
- Vitamins D, C, A, and E
- B vitamins (especially B_{12}, B6, and folic acid)
- Minerals, especially magnesium and zinc
- Alpha lipoic acid, coenzyme Q10, and pterostilbene with resveratrol
- Hesperidin, quercetin, curcumin, resveratrol
- Diosmin

Please refer to page 177 for more specific supplement and dosage recommendations.

processed grains and sweeteners. The Mediterranean diet is associated with lower all-cause mortality, decreased risk of metabolic and cardiovascular disease, and lower body weight and metabolic syndrome. The Mediterranean diet is typically lower in saturated fats and higher in oils, especially monosaturated olive oil. The polyphenols in olive oil have been shown to be especially beneficial in Alzheimer's disease and other neurologic disorders. Dr. Steven Gundry, proponent of the lectin-free diet, has stated that the only use of salad is to be a vehicle to deliver polyphenol rich olive oil. The diet also propounds the use of a moderate level of red wine with meals—no more than one to two glasses a day. Of course, it's important to recognize that along with the Mediterranean diet, exercise and stress modification are essential partners.

Ketogenic diet: Basically, the ketogenic diet substitutes and replaces sugars with fats as a source of energy. The advantage of this is that insulin isn't needed to deliver the energy to the cell. Reducing insulin reduces the risk of metabolic syndrome, heart disease, cancer, and a number of other disorders that are associated with insulin overproduction. Since excess sugar (the glucose) is pro-inflammatory, it is especially beneficial to the brain to reduce the level of sugar. The brain consumes 20 percent of the oxygen used in the body, yet is only 2 percent of our body weight, so there is a lot of metabolism in a small area. The conversion of sugar and the consumption of oxygen create a toxicity in the brain that sets up an environment conducive to brain disorders.

You may remember the term "ketoacidosis" in relation to a very serious complication of diabetes. And indeed diabetics must be very careful if they choose a ketogenic diet, but with a physician's help this can be effectively used. Ketogenic diet is the diet of choice for patients with convulsions (seizures). It consists each day of eating about 1.5 grams of protein for each kilogram of body weight and about 0.6 grams of complex carbohydrates per kilogram, with healthy fats, such as avocado, olives, olive oil, coconut oil, added to the protein and complex carbs. Conversion to a ketogenic diet is a surefire way of losing weight, and another way to develop mild ketosis is through intermittent fasting (see page 60).

Paleo diet: This diet is essentially heavy on meats, berries, and nuts, and resembles a keto diet. However, there is not enough research to suggest the long-term effects of this eating style, or to even know if that's how our hunter-gatherer forebears ate.

It is now thought that different groups of people respond differently to food. Recent DNA studies have shown us that there was much diversity in the ancient gene pool and different nutrition needs for them. This would seem to reinforce the idea that "one size does not fit all" in terms of the benefits of different diet strategies.

No matter what diet you prefer, you should emphasize fruits and vegetables, as there's an increased lymphatic flow from the polyphenols and flavonoids found in those. And because certain oils show a benefit to lymphatic flow, we prefer taking the Mediterranean approach for all the benefits, especially when it comes to lymph flow, preventing heart disease, weight loss, and more.

THE MAIN DIETARY APPROACHES, AT A GLANCE

The Diet	The Bottom Line
"Good Fats" Diet	The Mediterranean diet has been shown to lower risk of disease. It is a high-complex-carbohydrate diet with the addition of fish and small amounts of lean meat protein. Fruits and vegetables, legumes and grains provide high fiber and increased amounts of vitamin C. Fish supplies omega-3 EFAs, while olive oil offers monounsaturated fat. This diet was first described by Ancel Keys in the mid-1950s, and is most typically found in the Mediterranean countries of Italy, Greece, etc. Long-term studies have shown that those who adhere to the Mediterranean diet and lifestyle lower risk of developing diabetes, heart disease, and cognitive decline, and tend to live longer and age more gracefully.
High-Protein, Low-Carb Diet	Recommended by Atkins, Paleo and Keto diets also suggest eating vegetables, but limit fruits to berries and apples, and allow more meat, saturated fats, and dairy. They focus on no added sugars, as well as minimizing grains and beans. Effective keto diets limit protein, as well as carbs, and emphasize healthy fats. Excess protein gets made into sugars by the liver, which ultimately contributes to weight gain and loss of ketogenic diet benefits.
Very Low Fat Diet	Recommended by Pritikin, Ornish, McDougall, Gregor, Barnard, and others, this diet features high complex carbohydrates, high fiber, mainly vegetables, and 15 to 20 percent fat.

The Dietary Support System: Four Sneaky Things to Include in Your Diet

Butyrate: Butyric acid is created in the colon by helpful bacteria digesting a high-fiber diet. Butyric acid supplements have been shown to help create a healthy microbiome and create better gut-brain communication.

Small amounts of wine, coffee, cocoa, and dark chocolate: There's evidence to show that these are protective against the development of Parkinson's. The polyphenols have anti-inflammatory properties that are helpful, especially in older people with neurovascular issues.

Flavonoids: A form of polyphenols, they have a protective effect throughout the body. A Daflon supplement (the flavonoids diosmin and hesperidin) can help the liver, and is also protective of both blood and lymphatic vessels. They significantly increase lymphatic flow.

Polyphenols after a high-fat meal: They can help protect against blood flow restriction that happens after a high-fat meal.

CHAPTER 8

Fast Flow

How Exercise Improves
Your Lymphatic Fitness

In today's world, it seems that just about everything turns into a debate. That doesn't just apply to politics or seemingly innocuous Facebook posts ("You use garlic in your guac?! You're blocked!!"). The this-versus-that conversations aren't uncommon in the health and medical communities either. This can range from the serious questions (*Is surgery the best option?*) to the everyday ones (*How healthy is my daily IV drip of coffee?*). The truth is that when it comes to many medical issues, we probably also understand that there are nuances, subtleties, personal variations, and more gray areas than a smoggy skyline.

But there are also other health behaviors where the bottom line is as clear as a Caribbean lagoon. That goes for such things as "no smoking" and "brush, floss."

The other big one: exercise.

Physical activity, unquestionably, is a primary form of extra-strength prevention. And while "magic bullet" may be one of the worst phrases to use in the medical community (other than "your wait time is three hours"), we can say this: Exercise is such a powerful form of self-care that it sits atop the throne of good-for-you practices. Heck, if exercise were a job applicant, it would have one of the best-looking résumés of all, because it's been scientifically demonstrated to:

- increase lean muscle mass and strength
- elevate the basal metabolic rate
- help maintain weight loss
- decrease resting heart rate
- lower blood pressure
- lower low-density cholesterol (LDL, the damaging kind of blood cholesterol)
- increase high-density cholesterol (HDL, the protective form of cholesterol)
- lower blood sugar levels, helping to prevent and treat diabetes
- increase oxygen and blood supplies to your muscles and the arteries of your heart (coronary arteries)
- improve your overall heart function
- help prevent and treat heart attacks and strokes
- assist in the prevention of formation of clots in your blood
- help prevent and treat dementia, including Alzheimer's

That's why we need to move more, that's why docs constantly ask how much activity you're getting, and that's why exercise is a ubiquitous directive for many health issues, ranging from obesity and diabetes to cardiovascular diseases and cancer prevention.

Simple medical math: Moving, sweating, and pumping equals living longer and better.

It's likely that you already have some sense of why exercise strengthens your body in ways that few other things can. In this chapter, we're going to add to the long list of exercise's accolades by showing you how it relates to your lymph system—and how it serves as a flow-booster that will improve the overall quality of your lymphatics. And we'll show you the best way to move, in-cluding a look at the best lymph-friendly piece of equipment you can use.

Exercise and Your Lymph

You know the way the exercise culture works today. Slim tummy, six-pack abs, booty-boosting workouts. There's so much talk and marketing that centers around the "attractive" parts of the human body. And that's generally okay—because in most cases, a fit-appearing body can also be a healthy-functioning one.

So while most of you probably wouldn't put let-it-flow lymph in the same marketing bucket as bulging biceps, exercise—at its core—is really about what you can do to change your insides. Here are the ways that exercise influences the function of your lymphatics:

Increasing and improving flow: As you now know, the rate at which your lymph flows is a vital component for a healthy body. Research shows that exercise gets your lymph flowing at least three-fold—and in some cases, as much as seven-fold its resting level. Think of it working similarly to the "use it or lose it" concepts. When you don't use your muscles (say, if an arm or leg is in a cast), your muscle loses size and strength. You continually build a muscle's strength when it's working.

When you exercise, you're increasing the flow rate of your lymph, and that, in a "use it or lose it" cycle, keeps the fluid flowing. When you don't move regularly, the river of health becomes stagnant and still—and thus slows and decreases the overall function and flow.

Churn your body through a workout and you churn your lymph through your body.

This flow rate is also influenced by the way your muscles work during exercise.

As your lymph fluids are massaged along by adjacent muscle contractions, the one-way valves within the tubes of your lymph apparatus prevent the fluids from going backward. Whenever your muscles contract, they gently push the lymph fluid along. In addition, muscle contraction is one of the best ways to control the pressure of the lymphatic system, especially the core muscle of the diaphragm, which impacts the larger central lymphatic vessels that pass alongside of this muscle on their way to return lymph to the heart.

Clearing toxins: One of the benefits of better flow from more exercise is that you can better clear toxins from your body and thus reduce the risk of diseases and conditions that are linked to those toxins. One example: In a scientific study, subjects were injected with a low dose of an endotoxin (E. coli), and people who did hard exercise had lower levels of the resultant circulating toxin than those who went right to bed. Exercise cleared the toxin via the lymph.

This clearing of toxins also happens because the sweat that's generated via hard exercise carries them out. These toxins would otherwise clog the lymph if they weren't transported out of the body via sweat.

Think of your lymph as serving that street-sweeper function—taking debris from your organs, canals, and everywhere in between and moving them out of your body (via waste in a variety of forms).

Improving circulation: While you know that "we are what we eat," we're also what we *circulate.* Exercise helps not only get the nutrition to where it needs to be in your body, but it also allows the maintenance department of your lymph fluids to optimize the health of your arteries and the overall function of your cells. Exciting new research has shown that regular active exercise, along with dietary change, can even reverse atherosclerosis, the stiffening of our arteries and their blockages by plaques that develop with age and can lead to high blood pressure, heart disease, and strokes.

In general, you can expect to add two years to your life expectancy per one daily hour of active exercise you do per week. Lower rates of cancer, diabetes, Alzheimer's, and arthritis have been documented in people who exercise regularly.

Active exercise lowers cholesterol, especially important since it decreases the more dangerous LDL form and increases the garbage-truck-like healthful type, HDL. One way that exercise accomplishes this is simply by helping you lose weight, which itself lowers LDL. Active exercise increases levels of enzymes that move LDL out of the bloodstream, cells, and arterial linings into the liver for processing.

In this way, exercise works like a powerful conditioner of the lymphatic system. The role of the lymphatic rivers in chronic degenerative diseases such as atherosclerosis is often overlooked in typical therapeutic regimens. It is, however, important in the clearance of excess fats and toxic substances in your artery walls (which can contribute to chronic inflammation of the arteries). Exercise helps keep your lymphatic tubes clear, flexible, and optimally open.

Mounting scientific evidence demonstrates the part that the lymphatics play in arteriosclerosis, arrhythmias, and heart attacks, heart failure, and a host of symptoms related to chronic coronary insufficiency. HDL that enters from the channel of the artery exits the arterial wall by way of the adjacent lymphatics, and LDL does not. HDL enters the wall almost three times as quickly as LDL does, thus removing cholesterol from the innermost portion of the artery, then by way of the lymphatic tubes, taking this potentially damaging cholesterol to the liver, where it can be processed out of the body. Lymphatic fluid, which occurs in larger volume than blood, is completely circulated, on average, once every day. Half of your blood fats pass through your lymphatic system each day.

Because exercise can increase lymphatic flow by three to seven times as much, or more, it increases the clearance of cholesterol, and other blood elements, such as peptides and glycoproteins, from the arterial wall. It is simple arithmetic to see how by increasing your exercise, you can increase the circulation of your HDL in your bloodstream, thereby mobilizing more cholesterol from your artery walls. Experiments have shown that cutting off lymph is related to size of heart attacks (bigger heart attacks when lymph is cut off).

So it is evident that good lymphatic health can prevent or minimize the area of injury in your heart, as well as any arrhythmias.

Powering your brain: Your brain, right now, is probably working to make a decision about how you're going to approach exercise. What your brain decides, your body will follow. And that's a good thing, since exercise will also help your brain in so many ways.

For example, it boosts your mood, by changing not just the hormonal activity (you've heard of the runner's high?), but also the structures in your brain. And it has been linked to helping prevent memory loss associated with aging. Lymph channels have recently been found within the brain itself.

One of the ways active exercise helps your brain function may be through increased lymph flow within and from the brain itself.

The exuberance you feel after a session of exercise is due to several factors: an increase in the amount of endorphins (your natural opiates that are a hundred times more effective at relieving pain than morphine, but do not pass into the brain) and the natural activation of your endocannabinoid system, the same receptors activated by marijuana, without the short-term memory difficulties and long-term negative brain effects of ingesting this drug to get high. A fat-soluble substance called anandamide, present at increased levels in the blood after a session of exercising, *can* cross into the brain. The exhilaration you feel from exercise, along with diminished pain and anxiety, can last all day, minus the side effects and post-drug "crash" symptoms of marijuana use.

Active exercise has been shown to help your memory, including working memory (recalling recent events), spatial memory (remembering directions, and, for example, recalling where you are going while you are driving), as well as long-term memory (declarative memory). You grow more brain cells when you exercise regularly, a process called neurogenesis. It used to be believed that the number of brain cells stayed the same throughout your lifetime, but newer understanding has found many factors that can help you grow more brain cells, assisting with improved cognition, as well as recovery from strokes.

Even if you carry the genes associated with disease, active exercise can help change their expression. The science of epigenetics has demonstrated this can help all of your cells prevent and reverse the effects of any biological inheritance, along with the ravages of aging. This is shown through epigenetics, in which diet and lifestyle behaviors can change the way genes are expressed—meaning that behaviors can have an influence on proteins and peptides that are expressed, which send messages through the lymphatics. The implication: We have some influence on lymph flow, which can change the way our genes are expressed.

The Lymph-Boosting Workout

Without question, one of the biggest problems we have societally is that we spend more time on our butts than working them. "Sitting is the new smoking" is a common catchphrase in medical circles these days because of the health threat associated with being sedentary—the effect of spending most of our days at desks or driving or in front of screens. Risk of premature aging and dying at younger ages is elevated by how many hours a day we sit. Part of the reason behind this is lack of flow within your lymphatics. One Harvard study found that if you could reduce your time in the sit position to less than three hours per day, you could add two years to your life. Using a treadmill desk, an exercise ball–chair combination, or just standing and stretching a bit every half hour can help, but active exercise, half an hour at least three times a week, provides the best protection.

So at the very base of any health plan is to do anything to get moving. Ideally, we're shooting for at least three days a week for a minimum of thirty minutes (but more is great when you're ready). Before you get started, keep these things in mind:

- Make certain to get any exercise routine approved by your healthcare provider before you start, and then continue to be monitored as well. If you have become sedentary, are over forty, or have a history of heart disease in your family, it is wise to have a "stress test" of your heart prior to beginning any moderate physical activity, to help rule out the possibility of silent coronary artery disease.
- Prepare to start slowly, with supervision. It is well known that when a physician encourages patients to exercise, compliance increases. Try to find a healthcare provider who is familiar with the benefits of exercise. Extreme exercise is to be avoided; it increases your oxidative stress (harmful effects of lack of oxygen) and reduces levels of antioxidants and coenzyme Q10 in the body. Good cardiovascular tone requires only about thirty minutes of daily walking at 1½ to 4 miles per hour.

- If you're new to exercise, just start by making it a habit to get up and move every half hour. Walk around, stretch and breathe. Do a few jumping jacks, and wall push-ups, to ensure that your white cells get to where they need to go for their crucial maintenance work.
- If you have had a serious illness, such as severe cardiac disease, or have physical difficulty exercising, just aim for five minutes of walking daily to begin your program. You can imagine this in a chair, working your legs, marching in place, and then gradually move to walking upright with support. Add more as tolerated. As you watch TV, get up and stretch and walk a bit during commercials, or at least every half hour whenever possible.
- Take a walk before you eat. Walking before eating can offset the detrimental effects of high-fat meals, lowering by 25 percent the levels of blood triglycerides (a type of fat related to how much sugar you eat, above-normal levels can lead to heart disease).
- The most important focus of exercise has to be on regularity. If you can set aside some time in the morning, you are less likely to be interrupted; if not, you can do your session at home or go to the gym after work. Wait at least an hour after eating, and don't exercise too close to bedtime, as you may then take longer to fall asleep.

Once you slowly increase your fitness (or if you're already physically prepared and cleared for more rigorous exercise), you can try one of the following four methods (note that any type of exercise helps your lymph, as muscle contraction helps move the fluid, and also note that we'll cover yoga in depth in the next chapter because it's so crucial to lymphatic flow:

Cardio Exercise
(increasing your heart rate during activity)

Activities: Fast walking, jogging, bike riding, tennis, dancing, ellipticals and treadmills, stair climbers, group sports, and swimming

How to Do It: Whatever activity you like, do it at a pace enough to raise your heart rate (you should break a sweat, but it shouldn't be so intense that you can't maintain the pace for twenty or so minutes). We recommend thirty minutes three times a week to start (and you can break that thirty minutes into increments if that works best for you).

Why to Do It: Benefits have been shown to help prevent and treat heart disease, strokes, diabetes, obesity, depression, and Alzheimer's and other forms of dementia, and may even assist in prevention and help with the treatment of cancer. Brisk walking for three hours or more per week was found to be equivalent to other more strenuous types of exercising in reducing risk for coronary events (heart attack or death from coronary disease) by 20 to 30 percent in the women assessed by the Harvard Nurse's Study. Research in the United States and Japan showed a twofold decrease in mortality and complications of coronary artery disease in patients who regularly walked compared to those who did not. One study found that even low levels of walking in seniors is advantageous. Seven hundred nonsmoking men, ages sixty-one to eighty-one, were followed for twelve years. The mortality rate among those who walked less than one mile per day was nearly double of those men who walked more than two miles per day.

Path to Success: Some ways that it can work for you:

- Do it in a group (or have a walking buddy), or participate in a group activity, like a spin class. Community can help keep up your resolution to exercise regularly.
- Choose what you enjoy the most, not what you think is best. For example, if you're crunched for time, brisk walking may work best because no equipment is needed.
- Work around your limitations. If your joints hurt, you may find even walking difficult. Try swimming to relieve some of that pain.

Caveat:

- Exercise outdoors for the added benefit of more oxygen. (But beware of snow shoveling and other winter challenges: they put you at higher risk of heart attack because your fluids don't circulate as readily in the cold. Blood and lymph vessels constrict in cold weather.)

Feeling Adventurous? Pilates is a series of active muscle movements that combine some elements of strength training (see below) with the benefits of cardio. You can learn it via the internet for free; just make certain, as with all exercise, to start slowly and gently, so your muscles do not become too sore from being moved in new ways.

Strength Training
(*movements putting muscles to work*)

Activities: Some kind of routine where you are pushing and pulling (could be with weights, bands, or even your own body weight).

How to Do It: Pick three or four exercises (such as squats, lunges, etc.) and do ten or twelve repetitions of them at a time. You can do all these movements with dumbbells or resistance bands or even no weight at all as you lower and raise your own body weight. You can search the internet for a variety of movements from beginner to advanced.

Why to Do It: Well-toned muscle helps push the lymph fluids along. Lifting weights helps maintain muscle mass, prevents and treats osteoporosis, and optimizes metabolic rate, and helps prevent obesity.

Path to Success: If you're just starting out, try these methods:

- Strap wrist and ankle weights on while you do your daily active cardio exercising.
- Just doing sit-ups and push-ups can provide strength training, especially for beginners or travelers, since no equipment is needed. Wall push-ups are very simple: just place your hands on a wall and do a push-up. It's like you're doing them vertically rather than horizontally.
- Work out with a friend (preferably someone who has done it before) to help not only with motivation, but also with technique and ideas for new moves to try.
- Soup cans or jugs with handles filled with water or sand can be instant simple objects to lift.

Feeling Adventurous? Purchase a set of resistance bands, which allow you to do a variety of exercises and work your entire body. They come in a variety of resistances so that you can add more as you get stronger.

HIIT (High-Intensity Interval Training)
(*cardio exercise with more intensity*)

Activities: Any cardio exercise listed above but at a higher-level of intensity in short spurts, with lower intensity for recovery in between.

How to Do It: High-intensity interval training (also called SIT, sprint interval training, or sometimes just interval training) is a method of exercising with bursts of fast periods of one to three minutes, alternated with longer, slower recovery times. *(This should not be attempted if you have a known disease, and it is advised to begin only after being cleared by your doctor and coached by an experienced trainer.)* To begin, the usual routine is to first warm up, then do a few seconds of high intensity, going at maximum output, followed by a few minutes of recovery. Then you can gradually increase the time in high intensity, but still followed by full recovery back to your usual rate of exercise. The total session can be a few minutes at first, working up gently to twenty to thirty minutes.

Why to Do It: This approach has been shown to most quickly and effectively condition the body. And the high intensity is great for getting lymph movement.

Path to Success: You should be in good condition before you try this. Choose an activity that you like—and make sure to give yourself plenty of recovery time.

Feeling Adventurous? Group classes can really ramp up the intensity, combining cardio moves and strength moves in one high-intensity session. Other sports—like tennis or basketball—are also forms of HIIT.

Just be cautious not to overdo it; go on to other forms of exercise after a short session.

Mini-Trampoline (Rebounder)
(*jumping movements*)

Activities: Jumping on a mini-trampoline

How to Do It: A few minutes can be beneficial; don't overdo it, but pace yourself carefully as you get used to this type of action. Start by doing just a few minutes then include a daily session of five to ten minutes.

Why to Do It: Jumping on a mini-trampoline is a form of exercise that is particularly suited to your lymphatic system, and it has the advantage that you can do it at home or in your office. Bouncing on a mini-trampoline not only can put some extra fun into your exercise routine, but your fluids are given an additional push and pull from gravity. Lymph flow has been estimated to be increased fifteen to forty times as much. According to a NASA study, jumping on a trampoline is 68 percent more effective at burning calories than jogging.

Path to Success: If you purchase a trampoline, keep these things in mind:

- Some mini-trampolines have a bar attached, so that you can hold on to it while you bounce, which is especially helpful to beginners and older individuals.

- You should begin at your own pace, gradually incorporating more types of moves as you feel comfortable.
- Don't try running or jumping jacks at all if you have poor balance or are in any way disabled. You can still get benefit from a trampoline by sitting on it and bouncing up and down.
- You can also gain benefit by sitting in a chair and putting your feet onto the surface and beginning your bouncing program this way.

Feeling Adventurous? As you become more comfortable, you can try running in place. You can do jumping jacks while on the trampoline, for even more challenge and benefit. For example, you can start with your arms at your sides and legs together, jump up and come down with your legs spread and arms over your head. Over several weeks you might want to increase the number of jumps. Adding wrist and ankle weights, or using barbells, kettlebells, or a medicine ball, can improve your ability as a rebounder and be more effective in your overall conditioning. Gradually aim for a hundred jumping jacks daily in about five minutes. As you become more adept, you can incorporate HIIT into your routine.

CHAPTER 9

Smooth Flow

Yoga, Stress Management,
Massage, and Relaxation
to Improve Your Lymph

When it comes to yoga, you may fall into one of three camps: 1) You see people bending, twisting, stretching, and pretzeling, and think, "No way in the world is my body going to be able to do that; 2) You probably have heard of the many benefits of yoga but think you don't have the time (or are too type A) to breathe iiiiiiiin and oooooout; 3) You already regularly practice because you know how much better your body feels when you do.

If you're in the latter group, then we don't need to hard-sell yoga to you: You already know—and feel—the many health benefits of one of the world's oldest forms of self-care. But even if you've typically written off yoga as too slow/hard/confusing, you're going to want to stay with us.

That's because your lymph system loves yoga more than puppies love sneakers left on the floor. In the last chapter, we reviewed why exercise helps your lymph system—and you understand that those bouts of intensity are part of what helps generate the flow that's so important to lymphatic health. Yoga works in a similar (yet opposite) way—by gently getting the fluid to circulate. And as you know by now, that's essential to improving the overall function of lymph.

But here's the big thing: Yoga is probably the most plastic of all exercise forms in that you can morph it and shape it to whatever works with your body and your schedule. How so?

For one, you don't need expensive equipment (just comfortable, loose clothing and preferably a mat). You don't need a membership to a gym or yoga studio (though if you like, it is nice to do classes with a community). You don't need a lot of space (just a place where you can fully lie down and stand up). You don't need a lot of time (even a few minutes a day is better than no minutes).

And to address maybe one of your biggest concerns: Even if your hamstrings are as flexible as an oak tree or you can't even fathom how body part A stretches over body part B like that, that's okay. At whatever level—from beginner to yogi master—you can benefit from practicing yoga. Even if the only thing you've ever stretched is a T-shirt that shrunk in the dryer, you can work at your own levels, even if they seem simple at first.

So as you're thinking about your overall health—and overall function of your lymph system to prevent disease and improve your longevity—you'll want to embrace yoga as one of the protective shields in your arsenal.

How Yoga Helps Your Lymph

In the last chapter, we discussed how active exercise helps push and pull your lymph through the body. That chapter was about huffing and puffing. This yin-yang approach of including yoga is important because yoga—in its gentle, classical style—contrasts with active exercise and leads to *different* kinds of results and physical and mental benefits.

Here's how it works:

As you remember from our early discussion of the lymphatic

system, the system does not have a central engine or pump (in contrast to the heart, which works to pump blood throughout the circulatory system). However, because the lymphatic system operates through pressure—and in-and-out rhythm—it does need *something* to constantly "prime the pump," i.e., move the fluids along.

A major pressure pump: deep, diaphragmatic breathing.

This is why yoga stretches, yoga deep breathing exercises, relaxation techniques, and meditation are so crucial to healthy flow. All of them use deep, diaphragmatic breathing as part of the practice. Important note: Most people, unless they actively engage in this type of breathing through one of these methods, do not do this kind of breathing properly. Even if someone thinks they're doing deep breathing, it's likely a more demonstrative shallow breath, rather than really engaging the diaphragm.

Besides taking you through the biology of how yoga works, we're also going to show you how to deep-breathe correctly, and explain the physiology of each of the recommended yoga postures, deep stretching, deep relaxation, and meditation.

We know, we know. You're living a fast-paced life and barely have time to stop and smell your glass of rosé, so you're probably wondering how you're going to slow down even more.

But when you realize that slow action is just as vital to your lymph health as the fast action of exercise, we think you'll give a few of the moves that we will outline a try.

Muscles and Yoga: How It All Works

If your muscles had a mantra, they might have this one: "What doesn't kill us makes us stronger." Anatomically, that's one of the ways in which they work. When they're under stress—as the case when you lift something, be it a weight or your own body weight—

the structures of the muscle break down and get torn apart. Your body knows this, repairs the damage, and builds them back up—this time, better and stronger in anticipation that you'll do the damage again. This process is what builds all-important muscle in your body.

Now, if you took a microscopic look inside the muscle, you'd see this when muscles are moved quickly, as is the case with active exercise we discussed in the last chapter: You have receptors within your muscles called Golgi bodies, which may sound more like they belong in a wing in the Louvre but actually serve quite an important purpose in your muscle.

During active engagement of the muscle, the Golgi bodies send the signals to tighten and shorten the muscles themselves. This muscular activity is what develops tonicity (hence the term "muscle tone"). This process is what makes the muscles stronger and ready for you to lift your next weight or a feisty toddler. We need to maintain tonicity for strength, for metabolism, for lymphatic flow.

Yoga, on the other hand, essentially sends a shut-the-heck-up signal to the Golgi bodies, so that they do not engage. That's because the movements of classical yoga are performed slowly—meaning that instead of shortening and tightening, the muscle are lengthened and loosened. As a result, you feel more relaxed after a yoga session (not unlike the feeling after a massage).

So where does the lymph come in? This muscle movement pushes lymph along, as it squeezes the lymph through the channels. But it also plays a role in allowing your muscles to relax, with assistance from the relaxation hormones triggered by yoga. This relaxation is key because it assists the flow in absence of that central pump, not to mention that different body positions sort of jostle the fluid in different directions—allowing gravity to assist in the up-down path that lymph must take to be most effective.

As you relax and your muscles let go of their tension, more physical room is available for your lymph fluids to flow. This is both a direct effect of the muscles relaxing, plus the fact that when you are more at ease, there are positive physiological effects on your nervous and endocrine systems that expand your lymph vessels directly. The bigger these tubes are, the more free transportation within them is possible.

Think of that river of health—or for these purposes, traffic on an interstate. If a fast-moving highway moves from three lanes to one lane because of construction, everything stops. Everything just putters along as three lanes of vehicles try to fit into one. But when you go from one lane to three, it's smooth sailing—as the highway opens up. Yoga opens up the lanes for travel so that lymph can move freely and quickly.

As we discussed in the last chapter, active exercise has been shown scientifically to increase lymph flow, by dilating the lymph vessels, as well as by pumping the fluid along better, via the massaging effects of the muscles contracting and releasing adjacent to the lymph ducts. After the lymph fluid is squeezed forward, the one-way valves located in the lymph tubes prevent it from moving backward. That happens with a minimum of twenty minutes of exercise, but yoga has the added bonus in that stretches move the lymph throughout the whole body (as opposed to just the arms, legs, heart and lungs of active exercise).

Yoga as a Stress Beater

Even say the word "stress" and it might set your eyeballs on fire. That's because "stress" in our world has been associated with all of the things that cause us angst—from deadlines to long to-do lists to financial troubles or inane frustrations like the barista putting in two shots of espresso instead of three.

But if you peel back our modern framing of the word "stress" and think about what it meant in the original context of medicine, you'll see a much different picture.

Stress was first used to describe the action and reaction of any change in the human body.

Pretty cool concept when you think about it, huh? Every time you eat anything (a carrot or a carrot cake) is a stress, because your body is undergoing a change and reacting to it (obviously much differently between those two foods). A virus is a stress (that's probably bad), but exercise is also a stress (that's probably good, unless the running you're doing is from a crime scene).

So in technical and medical terms, a stress—especially a difficult one—invokes the "freeze, fight, or flight" response. You either leave the situation (flight), engage in it (fight), or stand there, evaluate the situation, and decide what to do (freeze). It's how our bodies react to any physical or mental change. Think about your own daily stressors: You either leave it, deal with it, or take some time to decide what to do.

This response is mediated through your sympathetic nervous system, a group of nerves located alongside your spine, in the mid-back, that connect with your adrenal glands and help regulate the release of the hormone adrenaline. Adrenaline, of course, is what allows us to be ready to assess things.

While all of this is happening, imagine your body in a biological washing machine—getting tossed, spun, and cycled, while all the dirt and grime gets mixed in with the cleaning agents. It's important to reduce the impacts of stress because stress increases inflammation, suppresses your immune system, and causes sleep deprivation. The result: Your body is in a topsy-turvy state, a mix of good stuff and dirty stuff. If you opened up this washing machine mid stress-cycle and peeked in before it was finished, you'd see these biological changes:

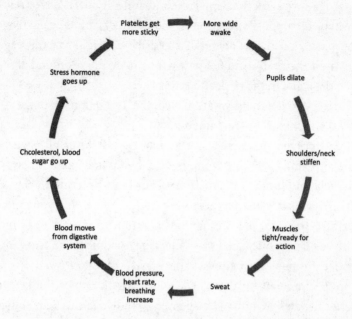

In the context of an evolutionary fight, there are very good reasons for all of these things happening: Your muscles and heart need to be ready for a fight, so they're engaged and ready to go; your pupils dilate to see your enemy better; and your platelets get sticky so that you don't bleed as much (in the event of a well-thrown spear).

But the downside is that in modern times, all of these stressors aren't needed to protect you against the eye darts your boss throws at you. Instead, they actually contribute to chronic disease, as chronic high blood pressure damages your circulatory system or chronic stress hormone increases your risk of belly fat, inflammation, and other problems.

What was necessary for your body in short-term stressful situations harms you in long-term ones. In addition, these stressors narrow the channels of your lymph system (by activating the sympathetic nerves and increasing adrenaline, as well as cortisol).

Yoga—as a mechanism for combating modern-day stress—invokes the opposite reaction in the nerve system. Yoga activates the parasympathetic nerves, located at its origin via the vagus nerve in the brain, and via parasympathetic nerves found alongside the spine in the neck and lower back.

This activation is what's called the relaxation response (or "tend and mend" or "rest and digest").

So the goal here isn't to eliminate stress. After all, stress is not something you can avoid. Besides being inevitable, stress is a medical and biological situation—your body reacting to change. The stress is designed to help your body—by helping you handle the stressor (like prompting you to action to meet a deadline). The key lies in our reaction—to be able to come down from high anxiety, to relax the pathways that allow lymph to flow, to let our body's systems calm down. Ultimately, that's one of the keys to better health and stress management—and yoga can help you get there.

Yoga for the Lymph: An Introduction

Although gentle yoga is generally quite safe, before you try the yoga recommendations in this book, be sure to get your medical practitioner's approval. In addition, follow these general cautions:

1. Go half as far as you think you can go.
2. Honor your own body.
3. If you strain, you will lose the relaxation benefits.
4. Practice regularly. Slow but steady.

Aim for pleasant actions, stay in your comfort zone, and you will gradually become more flexible and reap the rewards of practicing yoga without the risk of hurting yourself. Remember this motto: "The easy path is hard enough."

Any yoga practice generally helps move the lymph fluids along, but the poses we detail later are uniquely designed physiologically so you can achieve maximum lymph movement and overall relaxation.

Many schools of yoga are available; some methods are more aerobic than relaxing. To achieve optimum yoga practice, choose a type of yoga class that appeals to you and that is comfortable for you. But also do the special lymph sequence given below, adapted from the Integral Yoga tradition, to optimize your state of relaxation and your lymph circulation.

These yoga actions involve some stretching, but they are not just stretches. You stretch to come into a pose, where you let gravity do the work, as you relax your body into the position. Adapt the poses specifically to *your* body, to experience and maintain your peace, ease, and tranquillity. In this method, taking time to rest between actions is as important as the postures themselves, rather than flowing from position to position. This method of doing yoga gives you time to relax more, changing your lymph flow by the direct effects of this relaxation on the degree of muscle contraction, lymph vessel size, and nervous and endocrine effects.

You need both the movement of your muscles, to push the lymph fluids along, and then their letting go, to allow more space for subsequent flow, in order to move lymph fluids along most effectively.

If you make yoga a regular practice, you will see the following benefits, as the movements push along your lymph fluids directly and have nervous system and hormonal effects that impact rates and quantity of flow:

Better organ lymph: The structure of the lymph ducts, with their one-way valves, means that every time you move your muscles, you pump some lymph fluid along. Yoga poses, because they systematically target the various parts of the body, have a special benefit that regular exercise does not. Yoga can gently squeeze internal organs as well, and by turning the body upside down, as in the shoulder stand, move lymph along by gravitational as well as massaging effects.

Improved flow through stress reduction: Regular yoga practice makes your "fuse" take longer to respond to the stresses that come along in your life. Once you have experienced routine states of relaxation, you can maintain your relaxation more easily, or get back to it if you need. When we get tense or angry, for example, we tend to hold our breath. This not only raises blood pressure, which makes vessels more stiff, but it also has the effect of tightening muscles, preventing fluid flow, and impairing the lymph from getting to where it needs to go to do its work.

The Benefits of Stress Management

- Lowers the stress hormone cortisol
- Lowers blood sugar
- Lowers the formation of free radicals
- Increases endorphins
- Decreases incidence of cardiovascular disease

Improved nervous system: The central nervous system—the brain, spinal cord, and peripheral nerves—is in charge of the major elements of thinking, sensing, and moving. The adjacent autonomic nervous system has two elements: the previously mentioned sympathetic and parasympathetic systems, which control flight or fight responses. The vagus nerve is the main channel of

Massage and Flow

The countless benefits of massage range from reducing muscle soreness, easing stress, reducing fatigue, and improving exercise performance. Regularly reducing your stress by massage helps your body to maintain and repair itself internally, which assists in prevention and reversal of illness.

The other advantage: Massage can help your lymphatic system by improving overall flow. (There are even massages that specialize in lymphatic drainage, though any massage can help with this.) How does massage work? It reduces the workload of the lymph and the heart, making it not only a great therapy after any major health challenge, but also as a method of prevention.

In fact, you might consider physical touch as a kind of nutrient, in that skin-to-skin contact has been shown to be healing (research has shown that touch can reduce pain, blood pressure, anxiety—and even has been documented to reduce hospital stays in premature infants).

the parasympathetic system. *Vagus* comes from the Latin root meaning "wandering," and this large nerve wanders throughout the body. When stimulated, it relaxes muscles, blood vessels, and your lymph channels.

On page 181, we'll take you through a variety of poses and approaches so you can find a routine that works best for you.

You will have to experiment to see what style of massage you prefer, but here are some possibilities:

Lymphatic massage: Starts with light touch at the periphery and moves lymph toward the heart.

Swedish/deep tissue massage: Uses deeper strokes, with added oils and aromatherapy, and reaches the deeper, larger lymphatic tubes.

Alexander Technique: Combines posture training with light touch.

Rolfing: A deep-tissue technique that focuses on alignment and works with the fascia (connective tissue that surrounds muscles)— thought to be able to reverse scoliosis and carpal tunnel syndrome.

Therapeutic touch and Reiki: Practitioners lightly touch the body, or just move hands close to the body itself without touch. Healing is thought to be accomplished through energy exchange.

Reflexology and acupressure: Based on using pressure points on the soles of the feet, hands, ears, and other specific points, to encourage the movement of energy along "meridians" of flow.

Shiatsu: A kind of massage utilizing pressure points located at various sites on the body.

Thai yoga massage: Practitioners place the subject in various yoga positions, and massage them in such a way as to enhance yoga's flexibility and relaxation.

Note: You can easily learn simple ways to massage others and utilize self-massage as well. Simply use movements that go toward the heart, and this will move both blood and the lymph in the optimum direction. You can train your spouse, a friend, or your children, and trade massages.

CHAPTER 10

Well Meaning

The Connection Between
Spirituality and Lymph Flow

In health arenas, the traditional and most pervasive chatter re-
volves around the physical part of our bodies. We talk systems,
we talk cells, we talk about the biology about how part X interacts
with part Y and how it has an effect on Z. That, after all, serves as
the mechanics of medicine and the science of our systems.

In the last twenty years, the talk about the physical has been
supplemented by attention to the mind-body connections: the
role of psychology on wellness, and the intersection between
what happens between the ears and what happens in the rest of
the body.

But there's an important third area that gets about as much attention as your 398th favorite TV channel. That's the *spiritual* side.

This dimension—spirit and soul—has been shown to play a
key role in our optimum well-being. The spiritual aspects of our
lives have been documented by medical science to be vital to circulation, as well as affecting prevention and recovery from disease.
People who attend religious services regularly, for example, have
fewer heart attacks, recover more quickly if they do, and are less
likely to have another attack.

And that's really the way to think about total-body wellness: In
the three dimensions of physical, mental, and spiritual.

This applies to lymph flow as well. In fact, the great mystic, scientist, researcher, and teacher Emanuel Swedenborg, who is credited as first identifying the presence and actions of the lymphatic system around 1741, called cerebral spinal fluid "spirituous lymph." He understood that the flow of fluids within our bodies was essential for good health.

And in a way, you probably already understand that from what you have read in this book. Think about what you know about lymph now—how it flows, how it fluctuates, how it generates power (not through a central engine like the heart, but through variables that control the rhythm of the flow). That, in a sense, does have a spiritual and mystical quality to it, don't you think?

This chapter, however, isn't about identifying lymph as some kind of spiritual system; it's about how to use your own spirit to help improve your overall health. Depending on your perspective, background, and beliefs, this may be a difficult concept to think about: How, after all, could something we don't quite understand have a direct impact on tangible biological systems and processes? It's a deep concept, yes. After all, the body and brain are centered in time and space, but there is a part of the mind and whatever governs it that is "metaphysical," meaning it is not limited by these parameters of time and space—and even transcends them.

Now, we want to be clear: Soul and spirit not the same as religion. Soul and spirit (as you deduced) about who you are, how you feel, and the deeper meanings of existence, community, and connection.

While we don't want to imply that a grounded soul automatically means you'll have strong flow, we do think you should consider the importance of this third dimension as it relates to your overall health and wellness.

The Opportunity for Soul

Stress *hurts*, as anyone who has dealt with overbearing bills, crazy-uncle issues, bosses from hell, or any other life hurdles knows. Not only does stress make your mind feel like it's been chopped and churned in a blender, chronic stress also has dozens of negative impacts on your health.

As we have seen in this book, the burden of stress starts a cascade of chemical and hormonal changes in the body—namely through the hypothalamus and the adrenal axis—that cause inflammatory effects in the body. And that leads to a variety of tissue damage and disease.

Stress isn't just something you *deal* with. Stress is something that *attacks* us.

Now, in any situation like this, you can either avoid the attack or defend the attack. In life, it's impossible to avoid stress. Stress is simply a trade-off for the lives we live. So the question isn't how to avoid stress; it's what are the best ways to corral, manage, and defend against stress?

So how do you deal with the negative feelings that can result in painful physical consequences?

Sure, there are the day-to-day tactics that you may use to relieve the effects of stress—the bubble baths or sweat sessions, the deep breathing or foot massages, the glass of wine or cup of cappuccino.

But what if we take a look past the immediate fixer-uppers and think about a systematic approach to managing the ill effects of stress? Think about counteracting the damage through your own understanding, whether it be religious or personal. Think of what gives you assurance and a feeling of safety.

That is soul. That is spirit. That is self.

Soul could be considered as our being-ness itself, our pure awareness. Connections to our soul can be expressed and experienced most commonly in religious attitudes and activities, but it can also be achieved via spiritual practices, and ethical or moral principles. Whatever actions allow you to start out from a positive place rather than from emotional negativity can help keep you and your secret rivers safer.

The ability to use soul really starts with your initial defense against life's stresses.

After all, how we respond to stressful situations is our "response"-ability.

That can start with relaxation techniques, which help calm all of our systems.

The benefits of yoga practices and relaxation include making our "fuse" longer, and being able to choose positive pathways of reaction to a stressor, rather than anger, depression, anxiety, or the other negative emotions hurtful to our physiology.

Regular practice of yoga, massage, and other relaxing modalities give us a chance to observe the emotions we evoke in our daily lives. When we are already relaxed, we can say to ourselves: Do I really want to choose *that* emotion?

What can I substitute? Our spiritual understandings and connections give us new choices: compassion, empathy, forgiveness, kindness.

So it's not that yoga or relaxation makes the stress disappear. They give you the strength and emotional foundation to respond to and deal with the stressor—through actions that help your body, not hurt it.

What's most remarkable about cultivating soul solutions for our problems is that access to this remedy doesn't come and go with the inevitable roller coaster of life. Yes, in daily life we might get what we want, but if we are dependent on anything that "comes into" our lives from outside as a source of our happiness, it has a "going out" aspect, too.

Spirit and soul is by definition eternally there, available in all situations.

Spiritual practices can allow us to have an experience of peace and joy that is always within us ready to be recognized: our resting-peace self. "You were home all along, Dorothy," says Glinda the Good Witch to Dorothy at the end of *The Wizard of Oz*. L. Frank Baum, who wrote the books on which the movie is based, was a student of yoga, and wished to convey one of its essential truths: we have "home," source, pure consciousness, of who we are within us; meditation, prayer, contemplation, and mindfulness can help us carry this into every activity of life.

Thus, when we connect with Source, we can all become "Source-erors." Why should Harry Potter have all the fun?

Emanuel Swedenborg suggested we apply the idea that every thought and action can be assessed for the quality of its "use." This spiritual template can help us feel purpose, meaning, and peace even when things aren't going as planned. We can look for the silver linings playbook in anything, and instead of choosing negativity, we can find something positive to be of use to ourselves and others.

The Components of Soul

At first glance, it probably seems that soul is just something you are born with. You are who you are and that's that (sorry, that sounded like Dr. Seuss). And while there is some truth to our innate personality and emotional traits, the fact is that you *can* cultivate and develop soul and spirit. You can train your spirit. You can improve your well-being through development of soul. And you can think about your body—and your secret river of health—as something that is always being influenced by not only what you do, but also who you are. Here are some of the ways to think about how to get there:

Meditation: When you meditate, you change the functioning of your brain. Even the size and shape are altered. Areas associated with fear and stress, like the amygdala, become smaller, since they don't have to work as hard. The hippocampus, a region having to do with memory, increases in size, so that memory actually improves with age in those who meditate regularly. Here's what you can say to yourself when you feel like you just don't have time to meditate: Taking twenty minutes twice a day will not only ensure that your lymph fluids will circulate more optimally because you are more relaxed, but also your IQ will be boosted, so you can get whatever tasks you need to accomplish done more effectively!

To start, just try meditating for ten minutes. Find a quiet space, close your eyes, and just repeat a short phrase or mantra to yourself over and over. The idea is to clear your mind and block out all thoughts. (There are also many apps that can take you through the process if you like.) Just start with that short period and try to get into a daily practice. As you develop and see the benefits, you will likely start to increase your practice.

Mindfulness: Mindfulness is applied meditation. You take the quiet and peace of your formal meditation session into each action of your daily life. You watch whether you are clenching your jaw, hunching your shoulders, breathing fast and shallow with anxiety, or any other reaction to events. You then choose to consciously release the tightness, take a few deep breaths, and see if you can look at whatever is happening from a larger perspective of spiritual growth.

Many people associate mindfulness with how your body feels, and also in terms of eating. Mindful eating means thinking about every bite—and really engaging with all of your senses during your meals. This allows you to avoid mindless eating (i.e., lawn-mowering through a bag of chips), so you can be more in tune with eating healthy and the rich flavors that come with it. Slowing down when you eat helps you eat more consciously—and leaves you more satisfied. The same could be said about the way you approach a mindful day: Slowing down and being mindful of your movements helps you live more consciously—and leaves you more satisfied.

Prayer: While meditation is sometimes called listening to God, prayer is considered talking to God. Any form of prayer that keeps you relaxed may be helpful for your fluid flow. Even thinking of prayers during a busy day can help your secret rivers maintain their functions most appropriately. Grace before meals is a habit worth developing, both because it gives you a regular time for spiritual connection and contemplation, but also because invoking gratitude has been found to stop cycles of worry and depression.

Gratitude: Two of the most powerful words for your soul are "thank you." Research has shown that keeping a gratitude journal is one way to achieve these benefits: writing affirmations, reflecting on how much you have to be grateful for at the beginning and then also at the end of the day really works. Writing thank-you notes or emails also helps you feel good, as you acknowledge the kindness of others.

Laughter: Dr. Thomas Sydenham once said, "The arrival of a good clown exercises more beneficial influences upon the health of a town than of twenty asses laden with drugs." We are thirty times more likely to laugh if we are in a group setting (that's why laughter is inserted into situation comedies). It is often said that tragedy plus time equals comedy. Seeing the funny side of anything can help keep your secret rivers healthy as a form of exercise, and also as a way to connect to the underlying purpose and meaning given by a spiritual perspective. "All life's a stage, and we are just the players on it," wrote Shakespeare. Once we see ourselves in this light, we can identify with the bigger dance of the universe, and the meaning behind it.

Spiritual Love: Generally, our culture talks about love in a manner focused on individual relationships. Unfortunately, this can sometimes be the source of our stress: The challenges of differences, inevitable changes, and losses are everywhere (take a listen to a few country music song lyrics). A spiritual definition of love can give us a connection that endures, and also keeps our lymph flowing.

When we "fall" in love, the pleasurable brain chemicals dopamine, serotonin, and phenylethylamine (this last one is also increased by chocolate) are increased, and we want more. When we "rise" in spiritual love, we achieve and maintain a state of being happy from the inside out. We bring a happy, loving person *to* life, rather than expecting life to make us always happy.

"Alone" can also be spelled "all-one," meaning when we feel alone we can connect to the source of all of creation, through our spiritual practices.

Every time you choose spiritual love, even with your enemies, you keep yourself relaxed, and the internal fluids within your secret rivers can continue to flow. Yes, love may be the most important secret of all—and you can choose love to help yourself, as well as others.

Community: Loneliness has been found to be a risk factor for disease. "Loneliness is the poverty of the West, and it is a deeper poverty than the material poverty of the East," observed Mother Teresa. If we feel limited to our flesh-made container bodies, it's easy to feel disconnected and lonesome. We are social creatures, and thrive with adequate social support. (Social media, by the way, often increases our loneliness, and can be a trap full of negativity.) A notable exception: When St. John of the Cross was put in solitude in a jail cell because of his theological teachings, he used the time in meditation, and achieved a connection with God within himself that helped him survive until he was released.

A sense of belonging to a community larger than ourselves or our immediate family (although family is incredibly important) cannot be overstated.

So if you don't have one, you need to find your tribe—a small group of like-minded and supportive people who will give-and-take with each other as they journey through life's joys and struggles.

As Swami Satchidananda was fond of saying: "I" is the beginning of the word "illness," whereas "we" is the first part of the word "wellness."

A FLOW-FRIENDLY DIET

As you've seen, food can play a major role in your overall flow and health of your lymph system. To help you kick-start your eating approach, we've created a fourteen-day meal plan. It emphasizes high-quality foods and nutrients that can have a positive effect on flow, and is good for all of your body's system and functions. What's good for your lymph is also good for your heart, your brain, your weight, and your just-about-everything-else.

Once you get past fourteen days, go ahead and mix and match as you like. If you prefer one lunch rather than the others, then go ahead and make that the lunch you have most of the time. This is only intended as a guide for how you can include flow-friendly foods and develop a habit of healthy eating.

Week #1				
	Breakfast	**Lunch**	**Dinner**	**Snack**
Monday	Eggs and Leafy Greens Greek Scramble	Black Bean Burger over Greens with Avocado Dressing	Spaghetti Squash with Pumpkin Seed Pesto and Shrimp	Creamy Hummus with Carrot and Celery Sticks
Tuesday	Breakfast Quinoa Bowl with Apples, Ginger, and Cinnamon	Mediterranean Cauliflower Salad with Grilled Chicken	Thai Curry Bowl	Guacamole with Gluten-Free Tortilla Chips

	Breakfast	Lunch	Dinner	Snack
Wednesday	Almond-Oat Lemon Blueberry Muffins	Black Bean Burger over Greens with Avocado Dressing	Roasted Wild Salmon with Citrus Salsa	Easy Herb-Roasted Nut Mix, and an Apple
Thursday	Avocado Toast with Poached Eggs	Lentil and Chickpea Stew with Turmeric	Braised Chicken Tacos with Red Cabbage Slaw	Guacamole with Gluten-Free Tortilla Chips
Friday	Oatmeal with Raisins, Nuts and Apples	Chopped Asian Salad with Orange and Almonds	Spaghetti Squash with Pumpkin Seed Pesto and Shrimp	Creamy Hummus with Carrot and Celery Sticks
Saturday	Dairy-Free Berry Smoothie	Mediterranean Cauliflower Salad with Grilled Chicken	Spiced Chickpea Bowl with Roasted Tomatoes	Easy Herb-Roasted Nut Mix, and an Apple
Sunday	Eggs and Leafy Greens Greek Scramble	Lentil and Chickpea Stew with Turmeric	Braised Chicken Tacos with Red Cabbage Slaw	Creamy Hummus with Carrot and Celery Sticks
Week #2				
	Breakfast	Lunch	Dinner	Snack
Monday	Oatmeal with Raisins, Nuts and Apples	Black Bean Burger over Greens with Avocado Dressing	Ginger Vegetable Stir-Fry	Creamy Hummus with Carrot and Celery Sticks
Tuesday	Dairy-Free Berry Smoothie	Tuna Salad with Olives and Cucumber	Rosemary Baked Tofu with Asparagus and Gremolata White Bean Mash	Guacamole with Gluten-Free Tortilla Chips

	Breakfast	Lunch	Dinner	Snack
Wednesday	Breakfast Quinoa Bowl with Apples, Ginger, and Cinnamon	Lentil and Chickpea Stew with Turmeric	Halibut in Parchment with Citrus, Green Beans, and Tomatoes	Creamy Hummus with Carrot and Celery Sticks
Thursday	Avocado Toast with Poached Eggs	Tuna Salad with Olives and Cucumber	Braised Chicken Tacos with Red Cabbage Slaw	Easy Herb-Roasted Nut Mix, and an Apple
Friday	Oatmeal with Raisins, Nuts and Apples	Lentil and Chickpea Stew with Turmeric	Halibut in Parchment with Citrus, Green Beans, and Tomatoes	Guacamole with Gluten-Free Tortilla Chips
Saturday	Dairy-Free Berry Smoothie	Mediterranean Cauliflower Salad with Grilled Chicken	Spiced Chickpea Bowl with Roasted Tomatoes	Easy Herb-Roasted Nut Mix, and an Apple
Sunday	Eggs and Leafy Greens Greek Scramble	Apple and Shaved Brussels Sprout Salad with Spiced Walnuts	Braised Chicken Tacos with Red Cabbage Slaw	Creamy Hummus with Carrot and Celery Sticks

RECIPES

WEEK #1

BREAKFAST

Dairy-Free Breakfast Smoothie

Prep time: 5 minutes

Serves: 1

1 cup frozen raspberries
½ cup frozen cherries
1 cup frozen cauliflower
1 tablespoon almond butter
1 teaspoon flaxseeds
1 Medjool date, pitted
½ cup packed baby spinach
½ cup unsweetened almond milk
½ cup water
4 or 5 ice cubes, as needed

In a high-speed blender, blend all the ingredients until smooth and serve.

Breakfast Quinoa Bowl
with Apples, Ginger, and Cinnamon

Prep time: 10 minutes

Cook time: 15 minutes

Serves: 4

1½ cups quinoa, thoroughly rinsed
3 cups water
1 (1-inch) piece fresh ginger, peeled and grated
2½ teaspoons ground cinnamon
½ teaspoon sea salt
1 tart apple, peeled, cored, and chopped
1 cup unsweetened almond milk
1 teaspoon pure vanilla extract
3 tablespoons honey
2 tablespoons chia seeds
2 tablespoons chopped toasted pecans

In a medium saucepan over medium-high heat, bring the quinoa, water, ginger, cinnamon, salt, and apple to a boil. Reduce the heat to maintain a simmer and partially cover the pot. Simmer for 12 to 14 minutes, until the quinoa is cooked through and the apple has softened. Remove from the heat and allow to stand covered for 10 minutes. Fluff with a fork.

Add the almond milk, vanilla, honey, and chia seeds and stir to combine. Garnish with the pecans and serve.

Black Bean Burger over Greens with Avocado Dressing

Prep time: 15 minutes

Cook time: 30 minutes

Serves: 4

Black Bean Burger

4 tablespoons extra-virgin olive oil
1 small red onion, cut into small dice
3 garlic cloves, minced
2 teaspoons ground cumin
1 teaspoon smoked paprika
Pinch of cayenne pepper (optional)
2 (15-ounce) cans black beans, drained and rinsed
½ cup almond flour
¼ cup fresh cilantro leaves
1 large cage-free egg, lightly beaten
1 teaspoon sea salt
½ teaspoon freshly ground black pepper

Salad

1 ripe avocado, pitted and peeled
Juice of 1 lime (about 2 tablespoons)
3 tablespoons extra-virgin olive oil
1 (5-ounce) package mixed baby greens

¼ cup roasted and salted pumpkin seeds
1 cup fresh or thawed frozen corn kernels
1 cup cherry tomatoes, halved

Make the burgers: Preheat the oven to 400°F. Line a rimmed baking sheet with parchment paper and grease the parchment with 1 tablespoon of the olive oil.

In a medium sauté pan, heat 2 tablespoons of olive oil over medium-high heat. Add the onion and garlic and cook until softened, about 4 minutes. Add the cumin, paprika, and cayenne , if using, during the last minute of cooking. Set aside and allow to cool slightly.

Transfer the onion mixture to a high-speed blender and add half the black beans, the almond flour, and cilantro. Pulse until fairly smooth.

Transfer the mixture to a large bowl and stir in the remaining black beans and the egg. Season with salt and black pepper. Form the mixture into four ¾-inch-thick patties (using about ½ cup plus 2 tablespoons of the mixture for each). Place the patties on the prepared baking sheet and brush the tops with the remaining 1 tablespoon of olive oil.

Bake for 18 to 20 minutes, flipping the patties halfway through.

Meanwhile, make the dressing and assemble the salad: Wipe out the blender jar. In the blender, combine the avocado, lime juice, and olive oil and puree until smooth. Season with salt to taste.

To assemble the salad, divide the greens among four plates. Top with the pumpkin seeds, corn, and cherry tomatoes. Place a black bean patty on top and drizzle with the avocado dressing and serve.

Tip: To easily form the burgers, lightly wet your hands with water to prevent the mixture from sticking.

Mediterranean Cauliflower Salad with Grilled Chicken

Prep time: 20 minutes

Cook time: 25 minutes

Serves: 4

Cauliflower Salad

2 tablespoons extra-virgin olive oil

½ small red onion, thinly sliced

2 garlic cloves, minced

3 cups riced cauliflower

1 teaspoon ground turmeric

1 teaspoon sea salt

½ teaspoon freshly ground black pepper

2 cups baby kale

½ cup pitted Kalamata olives, chopped

2 Roma (plum) tomatoes, diced

½ English cucumber, diced

⅓ cup crumbled goat cheese

½ cup fresh mint leaves, chopped

¼ cup fresh flat-leaf parsley leaves, chopped

¼ cup sliced almonds, toasted, for garnish

Lemon Vinaigrette

2 tablespoons fresh lemon juice

¼ cup extra-virgin olive oil

½ teaspoon sea salt

¼ teaspoon freshly ground black pepper

Grilled Chicken

4 boneless, skinless organic chicken thighs
1 teaspoon sea salt
¼ teaspoon freshly ground black pepper
2 tablespoons extra-virgin olive oil

Make the salad: In a large sauté pan, heat the olive oil over medium-high heat. Add the onion and garlic and cook until softened, about 4 minutes. Add the cauliflower rice and turmeric and cook for another 5 to 6 minutes, until just cooked through. Season with salt and pepper and allow to cool slightly.

In a large bowl, toss the cauliflower mixture with the kale, olives, tomatoes, cucumber, goat cheese, mint, and parsley.

Make the vinaigrette: In a small bowl, whisk together the lemon juice and olive oil. Season with salt and pepper.

Make the grilled chicken: Heat a grill to medium-high or heat a grill pan over medium-high heat. Season both sides of the chicken with salt and pepper and drizzle with olive oil. Place on the grill and cook until charred and cooked through on both sides, about 3 to 4 minutes per side, or until a meat thermometer registers 165°F. Remove from the grill to a cutting board and thinly slice.

To serve, drizzle the vinaigrette around the rim of the cauliflower salad bowl and toss to coat. Divide the salad among four plates and top with the sliced chicken and toasted almonds.

Storage tip:

When prepping this salad for future days, store the cauliflower-onion mixture, kale, tomato-cucumber mixture, and vinaigrette in separate containers. Mix right before serving to keep the ingredients fresh and not soggy.

Tuna Salad with Olives and Cucumber

Prep time: 20 minutes

Serves: 4

3 (5-ounce) cans tuna packed in water, drained and flaked with a
   fork
¼ cup extra-virgin olive oil
2 tablespoons Dijon mustard
⅓ cup pitted Castelvetrano olives, chopped
⅓ cup fresh flat-leaf parsley leaves, chopped
⅓ cup fresh dill, chopped
⅓ cup chopped fresh chives
½ English cucumber, chopped
Juice of 1 lemon (about 2 tablespoons)
1 shallot, minced
1 teaspoon sea salt
½ teaspoon freshly ground black pepper
1 to 2 heads Bibb or Boston lettuce, leaves separated

In a large bowl, gently combine the tuna, olive oil, mustard, olives,
parsley, dill, chives, cucumber, lemon juice, and shallot. Season
with salt and pepper.

Serve the tuna salad on top of the lettuce leaves.

DINNER

Roasted Wild Salmon with Citrus Salsa

Prep time: 15 minutes

Cook time: 15 minutes

Serves: 4

Salmon

2 tablespoons extra-virgin olive oil

3 lemons, thinly sliced into rounds

4 (6-ounce) skin-on wild salmon fillets

1 teaspoon sea salt

½ teaspoon freshly ground black pepper

¼ cup fresh dill, chopped

¼ cup fresh flat-leaf parsley leaves, chopped

Citrus Salsa

1 navel orange

1 grapefruit

Juice of 1 lime

¼ cup chopped fresh chives

1 serrano pepper, minced (remove the seeds for less heat)

Make the salmon: Preheat the oven to 425°F. Drizzle the bottom of a glass baking dish with 1 tablespoon of olive oil.

Place four sets of two to three shingled lemon slices over the olive oil. Place the salmon fillets skin-side down and drizzle the top with the remaining tablespoon of olive oil. Season with salt and pepper. In a small bowl, combine the chopped dill and parsley and spread over the salmon fillets. Place the remaining lemon slices on top of the salmon fillets.

Bake for 12 to 15 minutes until the salmon is opaque and flakes with a fork.

Meanwhile, make the citrus salad: Cut a flat surface off the top and bottom of the orange. With the orange standing upright, move from top to bottom with the paring knife to remove the outer skin. Repeat with the grapefruit. Slice both fruits in half and then cut crosswise into smaller pieces. In a large bowl, toss to gently combine the chopped fruit with the lime juice, chives, and serrano pepper. Season with salt to taste.

Remove the lemon slices from the top of the salmon and serve garnished with the citrus salsa.

Spaghetti Squash with Pumpkin Seed Pesto and Shrimp

Prep time: 15 minutes

Cook time: 45 minutes

Serves: 4

Pumpkin Seed Pesto

⅓ cup raw pumpkin seeds

3 garlic cloves

1 cup packed fresh basil leaves

1 cup packed arugula leaves

1 tablespoon apple cider vinegar

2 tablespoons nutritional yeast

½ teaspoon sea salt

¼ teaspoon freshly ground black pepper

½ cup extra-virgin olive oil

Spaghetti Squash

1 (4-pound) spaghetti squash

2 tablespoons olive oil

1 teaspoon sea salt

½ teaspoon freshly ground black pepper

Shrimp

2 tablespoons olive oil

1 pound large shrimp, peeled and deveined

1 teaspoon sea salt

½ teaspoon freshly ground black pepper

½ teaspoon paprika

2 tablespoons chopped fresh basil, for garnish

Make the spaghetti squash: Preheat the oven to 400°F. Line a rimmed baking sheet with foil. Cut the spaghetti squash in half lengthwise and discard the seeds. Drizzle the insides with the olive oil and season with salt and pepper. Place cut-side down on the baking sheet and bake for 40 to 45 minutes or until tender when poked with the prongs of a fork. Remove and allow to cool.

Meanwhile, make the pesto: In a high-speed blender, pulse the pumpkin seeds, garlic, basil leaves, arugula, apple cider vinegar, nutritional yeast, salt, and pepper until roughly chopped. While the machine is running, slowly stream in the olive oil until smooth. To store the pesto, place in an airtight container with a piece of plastic wrap touching the surface of the pesto to prevent oxidation.

Make the shrimp. In a large sauté pan, heat 2 tablespoons of olive oil over medium-high heat. Season the shrimp with salt, pepper, and paprika. Place in the hot pan and cook until opaque, pink, and cooked through, 2 to 3 minutes per side.

Using a fork, shred the warm spaghetti squash into noodles and toss with ¾ cup of the pesto. Top the noodles with some of the shrimp, chopped basil, and serve with extra lemon if desired.

Thai Curry Bowl

Prep time: 15 minutes

Cook time: 30 minutes

Serves: 4

2 tablespoons coconut oil
1 small yellow onion, diced
2 garlic cloves, minced
1 (2-inch) piece fresh ginger, peeled and grated
1 teaspoon ground turmeric
2 tablespoons yellow, green, or red curry paste
1 zucchini, halved lengthwise and cut into half-moons
1 (15-ounce) can chickpeas, drained and rinsed
1 teaspoon sea salt
1 cup light coconut milk, shaken
⅔ cup water
2 cups sliced green beans (1-inch pieces)

To Serve

2 cups cooked cauliflower rice or brown rice
¼ cup roasted cashews, chopped
¼ cup fresh cilantro leaves
Lime wedges

In a large heavy-bottomed pot, heat the coconut oil over medium-high heat. Add the onion and garlic and cook until lightly browned, 4 to 5 minutes, then add the ginger, turmeric, and curry paste and cook for 1 minute more. Add the zucchini and chickpeas and cook for another 2 minutes, coating them with the curry paste and spices. Season with salt. Add the coconut milk and water and stir, scraping up any browned bits from the bottom of the pot with a wooden spoon.

Bring to a simmer, add the green beans, and cook for 12 to 15 minutes, until the green beans are almost tender.

Serve the curry over cauliflower rice or brown rice, garnished with the cashews and cilantro, with the lime wedges alongside for squeezing.

Braised Chicken Tacos with Red Cabbage Slaw

Prep time: 15 minutes

Cook time: 25 minutes

Serves: 4

Slaw

3 cups shredded red or green cabbage

¼ cup fresh cilantro leaves

¼ teaspoon ground cumin

Juice of 1 lime

2 tablespoons extra-virgin olive oil

½ teaspoon sea salt

¼ teaspoon freshly ground black pepper

Pulled Chicken

2 tablespoons extra-virgin olive oil

1 pound boneless, skinless organic chicken breasts, cut into
    2-inch pieces

1 teaspoon sea salt

½ teaspoon freshly ground black pepper

1 small red onion, diced

2 garlic cloves, minced

1 tablespoon tomato paste

1 tablespoon chili powder

2 teaspoons ground cumin

2 cups low-sodium chicken broth

Juice of 1 lime

To Serve

8 chickpea tortillas, warmed or charred
1 avocado, pitted, peeled, and thinly sliced

Make the slaw: In a large bowl, combine the shredded cabbage, cilantro leaves, cumin, lime juice, olive oil, salt, and pepper. Set aside to marinate while the chicken cooks.

Make the chicken: In a large heavy-bottomed pot, heat the olive oil over medium-high heat. Season the chicken with salt and pepper. Sear until golden but not cooked through, 7 to 8 minutes. Add the red onion, garlic, tomato paste, chili powder, and cumin and cook for another 2 minutes.

Add the broth and stir, scraping up any browned bits from the bottom of the pot. Partially cover the pot, bring to a simmer, and cook for 15 minutes, or until the chicken is tender and cooked through. Remove from the heat and shred the chicken using two forks. Stir in the lime juice.

Serve the chicken in the tortillas, topped with the slaw and sliced avocado.

SNACK

Guacamole

Prep time: 5 minutes

Serves: 2

1 avocado, pitted and peeled
1 shallot, minced
1 garlic clove, minced
2 tablespoons chopped fresh cilantro
Juice of 1 lime
Pinch of cayenne pepper (optional)
½ teaspoon sea salt

In a large bowl, mash all the ingredients using a fork or potato masher to the desired consistency and serve.

Creamy Hummus

Prep time: 5 minutes

Serves: 4

1 (15-ounce) can chickpeas, drained (liquid from the can
    reserved) and rinsed
⅓ cup tahini
2 garlic cloves
Juice of 1 lemon (about 2 tablespoons)
1 teaspoon sea salt
¼ teaspoon ground cumin
¼ cup extra-virgin olive oil, plus more for garnish
¼ teaspoon paprika
2 tablespoons chopped fresh flat-leaf parsley
Carrots, celery, cucumber, or bell peppers, cut into bite-size
    pieces, for serving

In a high-speed blender, pulse the chickpeas, tahini, garlic, lemon juice, salt, and cumin until finely chopped. With the machine running, slowly add the olive oil and blend until smooth. Add the reserved chickpea liquid as needed to adjust the consistency once blended.

Transfer the hummus to a bowl and garnish with additional olive oil, the paprika, and parsley. Serve with carrots, celery, cucumber, or bell peppers.

BREAKFAST

Almond-Oat Lemon Blueberry Muffins

Prep time: 15 minutes

Cook time: 25 minutes

Makes: 12 muffins

1 cup oat flour
1 cup almond flour
1 tablespoon ground flaxseeds
1 teaspoon sea salt
1 teaspoon baking soda
2 large cage-free eggs, lightly beaten
¼ cup date syrup or honey
1 teaspoon vanilla extract
1 tablespoon coconut oil, melted
½ cup unsweetened almond milk
Zest and juice of 1 lemon (about 2 tablespoons juice)
1 cup blueberries
½ cup chopped walnuts

Preheat the oven to 350°F. Line a 12-cup muffin tin with paper liners.

In a large bowl, whisk together all but 2 tablespoons of the oat flour, the almond flour, flaxseeds, salt, and baking soda. In a medium bowl, whisk together the eggs, date syrup, vanilla, coconut oil, almond milk, lemon zest, and lemon juice. Make a well in the dry mixture and pour in the wet mixture, gently stirring to combine.

In a small bowl, toss the blueberries and the reserved 2 tablespoons of oat flour together. Fold the blueberries into the batter. Fill the prepared muffin cups two-thirds full. Sprinkle the tops with the walnuts. Bake for 20 to 25 minutes, until a toothpick inserted into the center comes out clean.

Allow the muffins to cool in the tin for 10 minutes, then remove them from the tin and allow to cool completely on a wire rack.

Storage tip:

Store, covered, at room temperature for up to four days.

Eggs and Leafy Greens Greek Scramble

Prep time: 10 minutes

Cook time: 10 minutes

Serves: 2

2 tablespoons extra-virgin olive oil

1 shallot, chopped

2 garlic cloves, minced

2 cups baby kale

2 cups baby spinach

½ cup cherry tomatoes, halved

4 large cage-free eggs, lightly beaten

1 teaspoon sea salt

½ teaspoon freshly ground black pepper

2 tablespoons chopped fresh flat-leaf parsley leaves

2 tablespoons chopped fresh dill

2 tablespoons chopped pitted Kalamata olives

2 tablespoons crumbled goat cheese

In a large nonstick sauté pan, heat the olive oil over medium heat. Add the shallot and garlic and cook until softened, about 4 minutes. Add the kale, spinach, and tomatoes and cook until just wilted, another 3 to 4 minutes. Season the eggs with salt and pepper. Reduce the heat to low, pour the eggs into the pan, and stir gently, allowing the eggs to curdle slowly, 3 to 5 minutes, until just cooked through.

Remove the pan from the heat and sprinkle with the parsley, dill, olives, and goat cheese to serve.

LUNCH

Chopped Asian Salad with Orange and Almonds

Prep time: 20 minutes

Serves: 4

Salad

2 heads romaine lettuce, cored and chopped
1 bunch Swiss chard, leaves stemmed and chopped
2 navel oranges
½ cup chopped roasted almonds
1 cup sugar snap peas, cut in half
3 medium carrots, thinly sliced
1 red bell pepper, chopped
¼ cup fresh cilantro leaves

Coconut Aminos Dressing

2 tablespoons coconut aminos
1 tablespoon fresh lime juice
2 tablespoons rice vinegar
2 teaspoons honey
¼ cup avocado oil
½ teaspoon sea salt

In a large bowl, combine the lettuce and chard.

Using a paring knife, cut a slice off the top and bottom of the orange to create flat surfaces. Stand the orange upright and, working from top to bottom, use the paring knife to remove the outer skin and bitter white pith. Carefully run the knife between the segments of the orange to separate them from the membrane and add the segments to the lettuce mixture.

Add the almonds, snap peas, carrots, bell pepper, and cilantro to the lettuce mix and toss everything to combine.

Make the dressing: In a small bowl, whisk together the coconut aminos, lime juice, vinegar, and honey. While whisking, slowly stream in the avocado oil. Season with salt.

Drizzle the dressing around the rim of the bowl and toss until everything is lightly coated and serve.

Storage tip:

If you want to eat this salad throughout the week, store the ingredients in separate containers, then combine them and dress the salad right before serving.

Lentil and Chickpea Stew with Turmeric

Prep time: 10 minutes

Cook time: 30 minutes

Serves: 4

2 tablespoons extra-virgin olive oil

1 yellow onion, diced

1 red bell pepper, diced

3 garlic cloves, minced

2 teaspoons grated fresh turmeric

2 teaspoons ground cumin

1 teaspoon ground coriander

½ teaspoon ground cinnamon

Pinch of cayenne pepper (optional)

1 teaspoon sea salt

½ teaspoon freshly ground black pepper

1 cup dried yellow lentils, picked over for stones

1 (15-ounce) can chickpeas, drained and rinsed

4 cups low-sodium vegetable broth

1 cup water

3 cups stemmed lacinato kale leaves

2 tablespoons apple cider vinegar

In a large heavy-bottomed pot, heat the olive oil over medium-high heat. Add the onion, bell pepper, and garlic and cook until softened, about 4 minutes. Add the turmeric, cumin, coriander, cinnamon, and cayenne, if using, and cook for an additional minute. Season with salt and black pepper.

Add the lentils, chickpeas, broth, and water. Bring to a boil, then reduce the heat to maintain a simmer. Simmer with a lid on until the lentils are cooked through and the soup has thickened, about 25 minutes. Stir in the kale during the last 10 minutes of cooking and allow to wilt. Stir in the vinegar before serving.

Apple and Shaved Brussels Sprout Salad with Spiced Walnuts

Prep time: 15 minutes

Cook time: 40 minutes

Serves: 4

Walnuts

¾ cup walnut halves
1 tablespoon extra-virgin olive oil
¼ teaspoon ground cinnamon
¼ teaspoon paprika
¼ teaspoon dried rosemary
½ teaspoon sea salt

Apple and Brussels Sprout Salad

1 cup barley, rinsed
3 cups shredded Brussels sprouts
1 to 2 Honeycrisp apples, thinly sliced
½ cup fresh flat-leaf parsley, chopped

Dressing

1 small shallot, cut into small dice
2 tablespoons Dijon mustard

1 teaspoon pure maple syrup
3 tablespoons apple cider vinegar
⅓ cup extra-virgin olive oil
½ teaspoon sea salt
¼ teaspoon freshly ground black pepper

Toast the walnuts: Preheat the oven to 350°F. Line a rimmed baking sheet with parchment paper.

In a medium bowl, toss the walnuts, olive oil, cinnamon, paprika, rosemary, and salt together until evenly coated. Spread the walnuts over the prepared baking sheet. Roast for 10 to 12 minutes, tossing them halfway through. Remove from the oven and allow to cool.

Meanwhile, make the salad: Cook the barley according to the package directions and let cool. Transfer the barley to a large bowl and add the Brussels sprouts, apple, and parsley and toss to combine.

Make the dressing: In a small bowl, whisk together the shallot, mustard, maple syrup, and vinegar. While whisking, slowly stream in the olive oil to combine. Season with salt and pepper. Drizzle the dressing around the rim of the Brussels sprout mixture and toss until lightly coated. Set aside to marinate while the walnuts cool.

Toss the cooled walnuts into the salad and serve.

Tips: To prevent the apples from turning brown quickly, soak them in cold lemon water until you are ready to serve the salad.

If you are unable to find shredded Brussels sprouts at your local supermarket, buy whole ones (you'll need about 1 pound to make 3 cups shredded) and use the shredder attachment of your food processor to shred them.

DINNER

Spiced Chickpea Bowl with Roasted Tomatoes

Prep time: 10 minutes

Cook time: 40 minutes

Serves: 4

1 (15-ounce) can chickpeas, drained and rinsed
1 (12-ounce) container cherry tomatoes
2 tablespoons extra-virgin olive oil
1 teaspoon dried thyme leaves
1 teaspoon sea salt
½ teaspoon freshly ground black pepper
1½ cups millet, rinsed
½ cup fresh flat-leaf parsley leaves, chopped
½ cup fresh mint leaves, chopped
⅓ cup crumbled goat cheese

Lemon-Chive Dressing

Juice of 1 lemon (about 2 tablespoons)
¼ cup minced fresh chives
⅓ cup extra-virgin olive oil
½ teaspoon sea salt
¼ teaspoon freshly ground black pepper

Preheat the oven to 425°F. Line two baking sheets with parchment paper.

Pat the chickpeas very dry. In two separate bowls, in one, drizzle the chickpeas and, in another, the tomatoes with 1 tablespoon of olive oil each, ½ teaspoon of the thyme, ½ teaspoon of the salt, and ¼ teaspoon of the pepper. Place the chickpeas on one prepared baking sheet and the tomatoes on the other. Roast for 18 to 20 minutes, flipping them halfway through, until the chickpeas are crispy and the tomatoes have broken down and released moisture. Remove the sheets from the oven and allow to cool slightly.

Meanwhile, cook the millet according to the package directions. In a large bowl, fold the millet in with the parsley, mint, and goat cheese.

Make the dressing: In a small bowl, whisk together the lemon juice, chives, and olive oil. Season with salt and pepper. Drizzle the dressing over the millet mixture and toss to combine.

To serve, divide the millet among four bowls and top with the chickpeas and tomatoes.

Ginger Vegetable Stir-Fry

Prep time: 15 minutes

Cook time: 18 minutes

Serves: 4

2 tablespoons coconut oil

1 yellow onion, diced

3 garlic cloves, minced

1 (2-inch) piece fresh ginger, peeled and grated

½ bunch scallions, white and light green parts only, thinly sliced

2 cups small cauliflower florets

1 small zucchini, halved lengthwise and cut into half-moons

2 cups sliced baby bella mushrooms

2 cups small broccoli florets

1 cup sugar snap peas, cut in half

¼ cup coconut aminos

2 tablespoons apple cider vinegar

2 teaspoons honey

1 tablespoon sesame seeds

2 cups cooked brown rice, for serving

¼ cup fresh cilantro leaves, for garnish

1 lime, cut into wedges, for serving

In a large sauté pan, heat the coconut oil over medium-high heat. Add the onion and cook until lightly golden, 5 to 6 minutes. Add the garlic, ginger, and scallions and cook for 1 minute more. Add the cauliflower, zucchini, mushrooms, broccoli, and snap peas and cook, stirring occasionally, until lightly golden and softened but still with some texture, 7 to 9 minutes.

Meanwhile, in a small bowl, whisk together the coconut aminos, vinegar, honey, and sesame seeds. One minute before the vegetables have finished cooking, add the sauce and toss until it thickens slightly.

Serve the stir-fry over brown rice, garnished with the cilantro leaves and with lime wedges alongside for squeezing.

Rosemary Baked Tofu with Asparagus and Gremolata White Bean Mash

Prep time: 20 minutes

Cook time: 30 minutes

Serves: 4

Baked Tofu and Asparagus

1 (12-ounce) package firm tofu

4 tablespoons extra-virgin olive oil

1 teaspoon coriander seed, crushed

2 teaspoons sea salt

1 teaspoon freshly ground black pepper

1 teaspoon dried rosemary

2 (1-pound) bunches asparagus, tough ends trimmed

Gremolata

½ cup fresh flat-leaf parsley leaves, chopped

½ cup fresh mint leaves, chopped

2 garlic cloves, minced

Zest and juice of 1 lemon (about 2 tablespoons juice)

¼ cup extra-virgin olive oil

½ teaspoon sea salt

White Bean Mash

2 tablespoons extra-virgin olive oil

2 (15-ounce) cans cannellini beans, drained and rinsed

½ teaspoon sea salt

¼ teaspoon freshly ground black pepper

Preheat the oven to 425°F, with racks in the upper and lower thirds. Line two rimmed baking sheets with parchment paper.

Line a plate with paper towels. Place the tofu on top and place another layer of paper towels on top of the tofu. Weight down the tofu with a heavy pan and let stand for 10 minutes to press out excess liquid.

Cut the pressed tofu into eight ¼- to ½-inch-thick strips. Place them on one of the prepared baking sheets. Drizzle the tofu with 2 tablespoons of olive oil and sprinkle with the coriander, 1 teaspoon of salt, ½ teaspoon of pepper, and the rosemary. Rub the spices into the tofu. Place the asparagus on the second prepared baking sheet. Drizzle with the rest of the olive oil and season with the rest of the salt and pepper. Toss to lightly coat.

Roast the tofu on the upper rack for 25 to 30 minutes, flipping the tofu halfway through, until tender and golden brown with some crunch. Place the asparagus on the bottom rack during the last 15 minutes of baking the tofu and flip once halfway through.

Meanwhile, make the gremolata: In a small bowl, combine the parsley, mint, garlic, lemon zest, lemon juice, olive oil, and salt and allow to marinate until ready to serve.

Make the white bean mash: In a small saucepan, heat the olive oil over medium heat. Add the cannellini beans and mash gently using a potato masher, adding some water if necessary to smooth. Season with salt and pepper. Keep warm.

To serve, spoon the mashed white beans onto each plate. Top each with a drizzle of gremolata, some asparagus, and two pieces of tofu.

Halibut in Parchment with Citrus, Green Beans, and Tomatoes

Prep time: 15 minutes

Cook time: 15 minutes

Serves: 4

¼ cup chopped fresh chives

¼ cup extra-virgin olive oil

10 ounces haricots verts or green beans (ends trimmed, if using green beans)

2 cups mixed cherry tomatoes, halved

4 (6- to 8-ounce) skinless halibut fillets

2 teaspoons sea salt

1 teaspoon freshly ground black pepper

4 teaspoons fresh thyme leaves

2 lemons, thinly sliced into rounds

Preheat the oven to 400°F.

In a small bowl, combine the chives and olive oil. Cut four large heart-shaped pieces of parchment paper. Place a quarter of the haricots verts and tomatoes on one side of each heart. Top with a piece of halibut. Season with ½ teaspoon of the salt, ¼ teaspoon of the pepper, and 1 teaspoon of the thyme. Drizzle with 1 table-spoon of the chive oil and top with lemon slices. Fold the edges of the parchment tightly a couple of times to seal and form a packet. Repeat with the remaining parchment and ingredients to make four packets.

Divide the packets between two baking sheets. Bake for 14 to 16 minutes, until the fish is cooked through. Transfer each packet to a plate using a spatula. Open the packet just before eating—be careful, as the steam inside will be hot.

SNACK

Easy Herb-Roasted Nut Mix

Prep time: 5 minutes

Cook time: 30 minutes

Serves: 4

½ cup raw walnuts
½ cup raw almonds
½ cup raw pecans
½ cup raw cashews
2 tablespoons extra-virgin olive oil
1 teaspoon dried thyme leaves
1 teaspoon dried or fresh rosemary leaves
1 teaspoon fennel seed
1 teaspoon sea salt
½ teaspoon freshly ground black pepper

Preheat the oven to 300°F.

In a large bowl, combine the walnuts, almonds, pecans, and cashews. Spread the nuts over a rimmed baking sheet. Drizzle with the olive oil and sprinkle with the thyme, rosemary, fennel, salt, and pepper. Toss to lightly coat in the spices. Roast for 30 minutes, or until fragrant, stirring every 10 minutes. Remove the baking sheet from the oven and allow to cool completely.

Store in an airtight container at room temperature for up to one month.

SUPPLEMENT SUGGESTIONS

For those who aren't getting enough micronutrients, it's worth considering supplementation to make sure you get all the health (and lymph) benefits they provide. Here is our recommended list:

Supplement	Benefit	Recommended Dose
Omega-3 oils	These fats have many health benefits, including helping with the function of the brain, circulatory, and immune systems.	1 to 3 grams daily combined EPA and DHA
Vitamins D, C, A, and E	Vitamins serve many important functions, including protecting against disease. Some work as scavengers for damaging free radicals.	High-potency multivitamin and -mineral complex (as directed on label) Vitamin C: 500 mg twice daily Vitamin D3: 2,000 IU daily; check vitamin D levels regularly. If they are under 30, take 10,000 IU daily; over 50, take 2,000 IU daily. Take with vitamin K2. Vitamin E: 400 IU. Tocotrienols are part of the vitamin E (tocopherol) family, and have many benefits, including anticancer activity.
B vitamins	They serve a variety of roles, including to help the overall functioning of cells. They're also often linked to overall energy and health of the body.	Methylated B complex (as directed on label)

Supplement	Benefit	Recommended Daily Dose
Minerals, especially magnesium and zinc	Minerals provide many functions to help the body thrive. Magnesium works to maintain the function of nerves and muscles, while zinc is a key player in maintaining a healthy immune system.	Magnesium: Women 250–500 mg daily; men 500–1,000 mg daily. The most bioavailable forms include magnesium malate, citrate, lysinate, glycinate, threonate (complexed with the amino acids lysine, glycine, and threonine); threonine appears to carry magnesium into the central nervous system better than other forms. Zinc: Women 15–30 mg daily; men 30–50 mg daily. Gluconate and picolinate forms are well absorbed. Take with food.
Alpha-lipoic acid	This has antioxidant properties, which help to fight off disease by preventing damage to your cells. Equally good antioxidant in water-soluble and lipid-soluble compartments of the body.	300–600 mg daily, depending on level of oxidative stress in the body
Coenzyme Q10 (CoQ10)	This nutrient has an effect on metabolism, as well as disease prevention. Key compound in the mitochondria, the "energy factories" of every cell.	100–400 mg daily, ubiquinol form. Ubiquinol is the active (reduced) form of CoQ10 and is better absorbed, especially in older people. Blood level can be measured, and intake adjusted to optimize blood level.

Supplement	Benefit	Recommended Daily Dose
Pterostilbene	This polyphenol (found in many foods, but commercially isolated from the heartwood of *Pterocarpus* tree species) has similar disease-fighting capabilities like resveratrol, and is more bioavailable.	50–250 mg daily
Resveratrol	Found in grapes, this component has strong antiaging effects, helping to improve cell function and reduce inflammation.	50–100 mg daily
Quercetin	Found in many plants and foods, this plant-derived polyphenol helps protect against damaging free radicals.	250–500 mg daily. Better absorbed if combined with the enzyme bromelain, found in pineapple.
Curcumin	This polyphenol is found in turmeric. It has anti-inflammatory properties and provides strong protection for the gut, brain, and immune system.	500–1,000 mg, twice daily. Best absorbed in liposomal or nanoemulsion form.
Hesperidin	Found in citrus fruits, this bioflavonoid helps reduce inflammation.	50 mg daily

Supplement	Benefit	Recommended Daily Dose
Diosmin	This is a naturally occurring flavonoid found in plants (mostly citrus fruits). It helps reduce inflammation and promotes lymph drainage.	450 mg daily
Horse chest-nut seed extract	This extract promotes a healthy inflammatory response and supports the venous system.	250 mg daily
Grape seed extract	This polyphenolic mixture helps support collagen and elastin, which support blood vessel (and lymphatic vessel) walls, thus promoting better flow.	100 mg daily

BASIC YOGA
FOR LYMPH CLASS

There are many benefits to practicing yoga. Scientific research has shown that yoga can help prevent and reverse heart disease. Alternate nostril breathing has been shown to lower blood pressure. The practice of yoga, and its effect on lymph flow, has also been shown to help prevent and treat diabetes. Glucose tolerance and insulin resistance showed improvements after three months.

Below you'll find a complete class. We recommend you do it a few times a week (it would take about forty-five minutes to do the whole thing). But don't worry. If you don't have that much time, we have modified versions below—a twenty-minute version and a five-minute one. If you're a beginner, start slowly and ease into it. No matter how many minutes you do, you can often feel the effects right away.

So go ahead and find a quiet space with a yoga mat. You can do the practices in silence or pick whatever music feels soothing to you (we recommend *Healing Sanctuary* by Dean Evenson). Dim the lights, and be prepared to let your lymph flow.

For all yoga poses, you can find images and videos online by searching for the pose name. You may not be able to get into positions when you first start out. That's okay. The point is to get into whatever position you can—to help strengthen, stretch, and balance the body.

Humming, Chanting, or Singing

Begin the session by humming, chanting, or singing for a few minutes to help start relaxation.

Although music activates all regions of the brain, it shifts blood flow and electrical activity to the right side of the brain, to provide more balance between the left hemisphere of your brain (thinking, verbalizing) and the right hemisphere (nonlinear, whole-picture activity).

You can feel these right-left brain-balancing effects by a trial of humming, chanting, or singing right now. Whatever problem you are facing, try singing it to the tune of a children's song: "Oh, I'm having a hard time with _____, and I'm looking for some new solutions." Most likely, you will start to laugh. This is due to the brain looking at the situation differently by the action of this "music-mind," the name given to how your brain functions differently, with more right-left brain balance, on music.

The left brain measures linear time, and it's where we generally obsess about the past or future. The right brain does not experience time in the same way. We can free ourselves somewhat from the incessant stress of the left brain "worry mind" by activating the right brain "power of now" experience of this part of the brain.

Starting your yoga program with humming, chanting, or singing affects your lymph flow because as you are more relaxed, your lymph vessels expand. That's why the humming, chanting, or singing has to be a form that you like, and with which you feel relaxed. All you need is a few minutes. You can try traditional yoga chants like "Om" or "Om Shanti" (the *mm* sound can be very soothing; *shanti* means "peace"), or a hymn or song that you enjoy. You can do this silently, or skip this part of the class if you are not comfortable with it.

Scientific research has shown that humming, chanting, or singing helps lymph flow in several ways. As you are more relaxed, your lymph vessels dilate by the action of your parasympathetic nervous system, allowing more fluid to flow through.

In addition, when you chant the sound *mm*, such as in "Om" or "Om Shanti," the humming you do increases levels of a substance called nitric oxide, located in the lining of your vessels, which acts to make them dilate.

Eye Exercises

Eye exercises help by both pushing along lymph fluids and by relaxing you. This relaxation response allows your lymph vessels to dilate locally and throughout your body so that more fluid can flow easily through them.

In the embryo, the eyeballs and eye structures are created out of the brain tissue itself. Direct connections remain, and that's one reason you can often tell what someone is thinking by looking into their eyes, but not at their hand or foot. If you smile but your eyes do not, it can give away your true feelings.

Yoga eye exercises can reduce your stress level due to the connection of tension in your eye muscles to overall fatigue and tightness in your body. As your eye muscles relax, your whole body follows.

These exercises can even help your vision: In one experiment, the muscle-paralyzing agent curare was injected into the eye muscles, and the subjects' vision improved. This experiment demonstrated that tense eye muscles can affect the ability of the eyeball to adjust its shape whenever it needs to accommodate to visual tasks.

Yoga eye exercises should be performed in a way that is very soothing and relaxing.

End the yoga eye exercise session by massaging your closed eyes gently, from the inner portion to the outer, and feel your eyes and then your whole body relax. This massage directly moves the lymph fluids in the eyes.

Next, rub your palms together until you feel a sensation of warmth, then cup them slightly and place these "cups" over your eyes for a few moments. Feel your eyes relax further in the warmth and darkness. (See the book *The Eye Care Revolution* by Dr. Robert Abel if you want more information.)

The Sun Salutation

Most animals, after sleeping, start their day with some stretching. It feels good to start fresh and let go of any tension in the body that may have developed during dreaming. Stretching can also help you improve your lymph system function in a simple, joyous way.

The beauty of starting your yoga movements and postures with the sun salutation series is that this progression of poses warms your whole body and wakes you up, preparing you for the postures that are in the next portion of the class. If you practice this sun salutation set in the morning, it can help you have a better day. It gets your lymph juices flowing, so to speak, in a specific way to induce sequential effects—overall a wonderful way to greet the day, even if it's one when the sun itself is behind clouds. It can help you be in touch with your "inside sun."

The sun salutation sequence affects your entire body, squeezing your lymph fluid via the adjacent muscles' contracting in the various parts of your body while you go into and out of the poses.

- First, place your palms over your heart. You begin to feel your connection with the center of yourself. Taking such a prayerful attitude, if you wish, can also help you relax. We imagine this prayer gesture in history served to help any stranger feel safer, too ("Look, no weapons in my hands"), so that everyone could feel more relaxed.
- Next, with the thumbs locked, stretch upward toward the sky or ceiling and then very gently arch your back. Locking your thumbs allows you to pull a little, to more easily open up any stiff muscles, helping them make room for lymph fluids to flow. Make very sure not to strain. (Remember: The tortoise wins the race!)
- Next, slowly reach forward and try to touch your hands to your knees or the floor. Go only as far forward as you are completely able to stay at ease. Bend your knees, if necessary. Increased flexibility will come with more practice. Hold the

pose for only as long as is easy for you. Continue with the movements, as illustrated on page 184. Start with two or three sets; you may want to do more if you have time.

The sun salutation helps you warm up for the other yoga poses, and also helps you wake up *and* relax at the same time. The sympathetic nervous system, located in your mid-back alongside your central spinal column, is part of your nervous system that activates you by raising blood pressure, increasing heart rate, and narrowing the blood and lymph vessels directly. The backward-bending movement pushes on the sympathetic nerves and adrenals, which together also increase adrenaline, a hormone produced by the adrenals that helps you feel more alert, which triggers the freeze, fight, or flight response.

The forward motions of the sun salutation stimulate the para-sympathetic nerves, located at the base of the spine and neck, which put you in a more relaxed state. After you finish a few sets, you then begin to feel both more alert and yet more calm, ready to start your day. Regular practice makes this set of poses much easier and more natural. You won't want to start your day without it.

The joy of this practice is matched by its overall benefits. The backward bends have been shown to activate the sympathetic nervous system, and the forward bends affect the parasympathetic nerves, ending with you feeling uniquely alert, and yet calm.

The Upper Back Flex

This stretch activates your sympathetic nervous system, helping you wake up. In addition, it pushes the lymph along in the chest area, helping to protect your heart, and pushes the lymph along in the upper part of the body. It is especially important when considering the research showing the critical effects of lymph flow around the heart.

Lying facedown, place your hands underneath your shoulders, facing forward. Slowly lift up, but keep your elbows bent. Have easy breathing. Go only as far and hold only as long as is comfortable. As always, take time to relax between poses, as this helps with overall lymph flow.

The Lower Back Flexes

We sit way too much as a culture. This pose can help increase lymph circulation in your buttocks, providing relief to the commonly occurring "tight butt syndrome." These poses, done regularly, can help relieve low back discomfort and especially help prevent and treat issues with the discs between the vertebrae.

Lie on your back. After you put your arms underneath your body, palms down, you gently lift the legs, first one leg, then the other, and both together. Rest between each posture. As ever, be careful not to strain.

The Bow Pose

In this position, you can improve lymph circulation to your abdominal organs and help with pushing lymph fluid upward, where it can return to the blood circulatory system.

Gently assume the position by lying on your stomach and reaching back to have your hands meet your toes. Hold as long as is comfortable, then slowly rock back and forth. This action massages the lymph along, as well as the abdominal organs themselves, getting the lymph within them to move better.

This posture is especially helpful for diabetics, because it brings blood and lymph supply to the pancreas to optimize its repair work.

Forward Bends

Sit up straight and first stretch your right leg out, bending your left leg and placing your left foot wherever you can comfortably alongside the extended right leg. Lock your thumbs, breathe in, and slowly stretch upward, pulling on your thumbs for maximum reach. Arch your spine and while breathing out, bend forward from your hips, out over the right leg. Grab hold of your right leg, relax into the pose, and hold it only as long as is comfortable, with easy breathing. Just let gravity do the work. Repeat with the other leg, and then with both legs together.

These movements activate the parasympathetic nerves, located at the base of the spine, and thus make you feel especially relaxed and peaceful.

The Shoulder Stand

After having increased lymph circulation in both the upper and lower body, now you can use the effects of gravity to help drain whatever may have been stirred up so far. The shoulder stand is called the All-Members Pose, since it helps with the health of the whole body. This is a pose in which your legs are straight up in the air, with your shoulders on the ground, and your hands helping to support your body under your hips.

In this posture, gravity itself assists in returning the lymph fluid into the bloodstream, where it can then be processed through the liver, kidneys, and lungs. In addition, the position of the neck stimulates the parasympathetic nerves located there. Their activation invokes the relaxation response, and this directly enlarges your lymph vessels.

Come into and out of the position slowly. You can also place a pillow under your neck, or just lift your legs up against the wall in the modified pose. Hold only as long as you are completely comfortable.

You can gradually work up to staying longer in the pose, repeating it during the day. Shoulder stands lower blood pressure and relax the whole body, both by stimulating the blood pressure sensors in the neck (the baroreceptors) and, as noted, increasing parasympathetic nerve activity and relaxation, and thus increasing lymph flow.

In one study, when participants did a shoulder stand and then were exposed to the stresses of higher altitude, their thyroid function was shown to be improved, as compared to those who did not do a shoulder stand. Movement of lymph is a key part of this change.

The Neck Extension Pose

After you have flexed the neck forward in the shoulder stand, you can stretch your neck muscles in the opposite direction for one-third the time you have spent in the shoulder stand. It helps to relax the neck further after the effort of the shoulder stand, and also assists pushing along the lymph in the neck. Come into and out of the posture slowly and gently and, if needed, use a pillow to stay comfortable.

As you increase the time in the shoulder stand, you can stay longer in this follow-up pose, but only aim for one-third the time you spend in the shoulder stand. This pose releases stress from the neck and pushes lymph along.

The Half Spinal Twist

Now that you have drained the lymph fluids via the shoulder stand, you can further push the lymph through the kidneys and adrenals with this pose. It gently massages these organs, located in the mid-back.

One-quarter of every heartbeat is filtered through the kidneys, which are working hard all day to keep your blood balanced and healthy. Direct massage allows your lymph to get to where it needs to go in order for it to help the kidneys achieve their best function.

The adrenals sit like small woolen caps on top of the kidneys. The adrenals and kidneys are overactivated by the excess stress our culture routinely invokes in us. Spinal twist is especially helpful to women after menopause, when the adrenals can produce just enough estrogen to relieve menopausal symptoms, with none of the risks associated with estrogen replacement therapy.

Come into the pose slowly and gently, maintaining your comfort level. Hold only as long as you can easefully stay relaxed.

Crossed-Leg Forward Bend

Cross your legs comfortably, then sit up straight, take a deep breath, and, as you exhale, gently fold forward from the hips. Make sure not to strain; just relax into the posture.

Sometimes called the "Seal of Yoga," meaning it seals in the energy of relaxation of all the preceding postures, this pose helps quiet the body by stretching the parasympathetic nerves at the base of the spine, activating your relaxation nervous system. Hold only as long as is comfortable. Slowly sit up again, breathing in as you come up, and then sit for a moment, feeling the relaxed state achieved.

Yoga Deep Relaxation

The progressive deep relaxation is the dessert of the yoga class. It takes advantage of the mind-body connection. It is similar to when you stretch a rubber band, then let it go. Initially, you take a deep breath, tighten your muscles systematically, one by one, and then let them go, while making a whooshing sound through your open mouth.

You, like many of us, may have been carrying more tension in your muscles than you realized. Once you experience the effects of physically tightening and then letting go, you can say to yourself: "Oh, *THAT'S* what it feels like to really be relaxed!" Next, you focus your mind on the various parts of your body, starting with the toes and moving up the body, and direct these areas to relax even further. Your mind is remarkably capable of connecting to your body. For example, if sensors are placed on your thumb, and instructions given to make the thumb hotter by "telling it" in your mind to do so, and imagining it, in about six sessions, most people can make their thumb get warmer.

After you stroll through different parts of your body in your mind, focus your mind on your breath. Since breath and mind are intimately linked, as you slow the breath, you become more calm and relaxed. This is a skill you can use to help reduce stress, as we shall see in the section on breathing exercises below.

Next, bring the mind's focus to the space between the eyebrows. In the yoga tradition, this is the center where the body and mind connect to the soul, and it is called "The Third Eye." You just watch your thoughts, like clouds that come and go, while becoming even more peaceful. Finally, let yourself go beyond your changing thoughts, and experience the deep relaxation of your soul, defined as your essential pure consciousness.

Just rest in that deep peace as long as you like.

When you are ready, you can deepen your breath, feel it traveling through your body to wake it up, starting with the feet and hands. Begin to wiggle your hands and feet, then roll your arms and legs, and, still feeling that deep state of ease, slowly sit up, ready for the breathing exercises.

Taking time to do this deep relaxation once or twice daily can profoundly affect your lymph flow. The relaxation's effects on your nervous and endocrine systems enlarge your lymph vessels and prepare your muscles to relax enough to allow more flow of fluids within these tubes.

Visualization and Imagery

Once you are fully relaxed, you can spend extra time seeing, in your mind's eye, your body's fluids, nutritional resources, and immunity factors, such as white cells, flowing to where they need to go to reverse any diseased areas. You can imagine your white cells chomping down on areas of blockage or removing inflammation. Spend a few minutes doing this, using whatever imagery appeals most to you.

The Yoga Breathing Exercises

All that you have done so far prepares you for the even more profound relaxation effects of the yoga breathing practices. When we are feeling stressed, we breathe fast and shallowly; when relaxed, slowly and deeply. If we *make* ourselves take slow and deep breaths, we can directly affect our state of relaxation to a remarkable degree.

This works through something called the Hering-Breuer reflex. Stretch receptors in the lungs are activated when you take a deep breath, sending a signal to the brain via the vagus nerve to trigger parasympathetic nerve activity. This causes your whole body to experience the relaxation response. So even one deep breath instantly relaxes your whole body! Try it now: Take a deep breath and see how you quickly feel a bit more relaxed all over. Yoga breathing exercises are the most portable of all the yoga relaxation techniques; you can do them anywhere, whenever you are feeling stressed.

To get the deepest breath, three-part deep breathing is recommended. When we are stressed, we often breathe fast and shallow, from the chest. We can achieve more activation of the Hering-Breuer reflex if we use more diaphragmic action.

Babies naturally breathe this way; watch them asleep, and you will see the abdomen rise and fall with each breath in and out. Opera singers are trained to always breathe like this to maximize their singing capacity.

- Place one hand on your abdomen and the other on your upper chest. As you slowly breathe in, feel your abdomen expand to meet your hand placed there. Then expand your lower chest, and finally your upper chest. You are filling your lungs like a pitcher, from the bottom to the top.
- Slowly exhale from the top to the bottom, and finally give a gentle squeeze to your abdominal muscles, pushing out as much air as is comfortable.
- Continue this method of breathing, breathe in again and *slowly* repeat for as long as is comfortable. Just a few minutes are fine initially; you can gradually lengthen the time and repeat more than once a day for the most effective relaxation results. After you have gotten used to it, you do not have to use your hands to help.

After practicing this form of breathing regularly, you can begin to have abdominal breathing as your constant way of exchanging air *all day long*, rather than breathing just from your chest.

Following a few rounds of three-part deep breathing, do this alternate nostril breathing exercise. Air moving through the right nostril stimulates the sympathetic nervous system. When we breathe through the left nostril, it activates the parasympathetic nervous system via the vagus nerve. Thus, alternating between the two nostrils acts to further relax you by balancing sympathetic with para-sympathetic action.

In addition, within a short time, this practice balances right- and left-brain activity, changing our consciousness to a more relaxed state.

- Begin by using your thumb to close off your right nostril, and breathe out through the left. Utilize the three-part deep breathing technique, described on page 193, to achieve maximum air flow.
- Breathe in through the left nostril, again using three-part deep breathing.
- Next, breathe out through the right nostril; then in.
- Finally, breathe out again through the left nostril.
- Continue breathing through alternating nostrils for a few minutes.

You can gradually increase the time in which you do alternate nostril breathing, and add more than one session per day, to assist with the best relaxation, and thus flow of your lymph. Alternate nostril breathing has been shown to decrease blood pressure.

Breathing exercises help to prepare you for the harder work of quieting and directing your mind through meditation and mindfulness. They can also be used to help you feel connected spiritually, which can help reduce your stress. The word for "God" in Hinduism is *Brahman*, from the Sanskrit root *bri*, meaning "to breathe." This tradition postulates that God breathes out the evolution of the universe from his essence, then, when the world gets too crazy, breathes it all back in, like at the end of the childhood game of hide-and-seek, where the signal is given to start the game anew, "Ollie, ollie, oxen free."

Meditation

Because it is so powerful, meditation is the aim of all the previous yoga practices, which are meant to prepare you for this deeper work. We think about three hundred to a thousand thoughts per minute. Each one of them affects our stress level, and thus our body and its flow of lymph.

We can begin to slow down these thoughts, as well as replace more difficult thoughts with more peaceful ones, via meditation, to reduce stress directly. In addition, when the mind is more still, the deep peace of *our essential* consciousness itself can become more apparent. This affects our lymph flow: The more relaxed we are, the more the channels remain larger and the lymph fluids can flow.

- First, assume a comfortable position. You can sit on a chair, or on a cushion. Try to keep your back as straight as is comfortable.
- Next, choose a focus. You can repeat a sound, such as "Om" or "Om Shanti" (*shanti* means "peace"), "Relax," or "Peace"; you can use a short prayer; or simply observe your breath.

Start with a few minutes and gradually increase to twice a day, as long as you are completely relaxed. Most people find twenty minutes twice a day to be most useful in the beginning; you can increase as needed and comfortable for you.

The Twenty-Minute Yoga for the Lymph Routine

Whenever possible, doing the full yoga routine given above is recommended. However, if you can't fit it all in, you can do the basic essentials in a twenty-minute format.

- Start with three sun salutations, described on page 184, to warm up your muscles, help realign your posture, and give maximum push to your lymph fluids.
- Do the upper and lower back stretches to improve lymph flow in these critical areas.
- Turn the body upside down in the shoulder stand, so that gravity can help your lymph system to drain. Flex your neck backward afterward.
- Do a quick deep relaxation, tensing and releasing your muscles, helping to further move the lymph along, as well as relax the lymph vessels themselves. When the lymph vessels are wider, more fluids can move through them.
- Do a few deep abdominal breaths, which pushes against the largest lymph vessels in your chest, assisting circulation of lymph.
- Spend a few minutes in meditation, just focusing and relaxing your mind.
- Mindfully carry that state of relaxation with you throughout the day.

The Five-Minute Quick Yoga for the Lymph Program

- Do one round of Sun Salutation, described on page 184.
- Do a very quick deep relaxation, as detailed on page 190, but this time take a deep breath, tense all your muscles at once, then let your breath out with a gush, while relaxing all the muscles.
- Spend a few moments with three-part deep breathing, followed by a few minutes in meditation.
- Carry this renewed state of peace with you mindfully. Repeat this five-minute sequence as needed.

Acknowledgments

First, I'd like to thank my partners in writing this book, Drs. Dwight McKee and Sandra McLanahan, without whose help, knowledge, and experience, this book would not have been written. Next, kudos to Ted Spiker, a true wordsmith who transformed our medical palabra into a readable, enjoyable text.

I must also acknowledge my brilliant and beautiful wife, Emily Jane, who is the true motivator behind this book and who reviewed the diet, menu, and supplement sections of this book and made them whole. And my granddaughter Daphne Oz, who helped her grandmother, and to my son-in-law Dr. Mehmet Oz for his medical and practical acumen and assistance.

Many thanks to my administrative assistant, Eileen McKiernan, who facilitated transactions, typing, and production efforts; Christine Smythe and Doris Odhner for their splendid proofreading; Chris Conway, who helped with the supplement regimen; Nadia Chen for her illustrations; and Laura Arnold for her work on the menu. And last, the team at Scribner, starting with Roz Lippel for not only picking out just the right words but ferreting out the wrong ones. Many thanks also to Rebekah Jett, Katie Rizzo, and Mia O'Neill.

Index

Abel, Dr. Robert, 184
acupressure, 128
adipokines (adipocytokines), 59
adipodren, 42
adiponectin, 59
Alexander Technique, 128
Almond-Oat Lemon Blueberry Muffins, 160–61
alpha-synuclein, 73, 74, 76
Alzheimer's disease
  beta-amyloid plaques in, 68, 70–71, 73, 76
  complicated nature of, 64
  diet in prevention of, 72, 79, 96
  exercise and, 102, 105, 110
  genetic component of, 70
  glia cells in, 67
  increase in deaths from and incidence of, 64
  inflammatory markers in, 71–72, 73
  lymph-flow cleansing process in, 76, 77
  people's fears about, 63
  protective effect of resveratrol and pterostilbene in, 68
  tau protein in, 73
anandamide, 106–7
American Heart Association, 86

amyloid plaques
  Alzheimer's disease and, 68, 70–71, 73, 76
  exercise and, 104–5
antioxidant nutrients, 27, 42, 108, 178
Apple and Shaved Brussels Sprout Salad with Spiced Walnuts, 166–67
arteries
  blood glucose and plaque formation in, 57
  circulatory system with, 17–18, 22
  ways of improving flow in, 21–27
arteriosclerosis
  diet and, 86
  lymphatics' role in, 105
avocado recipes
  dressing, 145–46
  guacamole, 158

Baum, L. Frank, 135
BDNF (brain-derived neurotrophic factor), 72, 79–80, 81
beta-amyloid plaques, in Alzheimer's disease, 68, 70–71, 73, 76
beverage pyramid, 91

Black Bean Burger over Greens with Avocado Dressing, 145–46

Blaylock, Dr. Russell, 72, 75

blood-brain barrier
anatomy of, 67–69
brain inflammation from toxins passing through, 71–72, 75, 77

blood flow
circulatory system and, 16–17
exercise and, 104–5
lymph flow connected to, 22
ways of improving, 21–27

Blueberry Muffins, 160–61

body
caloric energy for, 84
epigenetics and functioning of, xviii–xix
flow-based systems in, 1
food's effect on, 85
lymphatic system's maintenance of, 3
stress cycle in, 122–23

bowl recipes
Quinoa Bowl with Apples, Ginger, and Cinnamon, 144
Spiced Chickpea Bowl with Roasted Tomatoes, 168–69
Thai Curry Bowl, 154–55

bow pose, in yoga, 187

brain
blood-brain barrier in, 67–69
gut connection to, 44–47, 74
MSG's impact on, 71–72
neuron communication in, 65–67, 74
overview of anatomy of, 64–69
parts of, 65

brain-derived neurotrophic factor (BDNF), 72, 79–80, 81

brain health
BDNF stimulation and, 79–80
diet and, 47, 78–79, 81
exercise and, 106–7
flavonoids and polyphenols for, 78–79
glymphatic system and, 10, 73, 75–76
lymphatics' role in, 64, 72–73, 76–78
meditation and, 136
music and, 181–82
new neuron growth and, 81
obesity's influence in, 55
sulforaphane and, 72
ways for improving lymph flow in, 78–80

Braised Chicken Tacos with Red Cabbage Slaw, 156–57

breakfast recipes, 143–44, 160–62

breathing
lymph flow and, 4, 23, 24
yoga and, 119, 181, 192–94

Brussels Sprout Salad, 166–67

butyrate and butyric acid, 47, 80, 81, 100

caloric energy, 84

Campbell, Colin, 87

cancer
diet and, 86–87, 89, 90, 97
exercise and, 105, 110
immune system response to, 32–34
lymphatic sluggishness and growth of, 38
lymphatics' role in changes in DNA expression in, 8
lymphatic system and, 36–39

lymph nodes and, 5, 37, 38–39
MSG as trigger for, 72
obesity and, 57
preventive methods for, 30
supplements for fighting, 37
capillaries, in circulatory system, 16, 17–18, 19
carbohydrates
  high-carb diets with, 91–92
  ketogenic diet and, 97
cardio exercise, 110–11
cardiovascular disease. *See* heart disease
cardiovascular health. *See* heart health
cardiovascular system
  hardening of arteries in, 7, 8
  lymphatics' role in disease and, 8
  overview of anatomy of, 16–18
  role of lymphatic system in, 9
  ways of improving lymphatic flow in, 21–27
celiac disease, 50, 95
central nervous system
  lymphatic tissue in, 76
  yoga and, 126
chicken recipes
  Braised Chicken Tacos with Red Cabbage Slaw, 156–57
  Mediterranean Cauliflower Salad with Grilled Chicken, 147–48
chocolate, in diet, 80, 81, 94, 100, 138
cholesterol
  exercise and, 105–6
  high-density and low-density lipoproteins and, 9, 18–19, 102, 105
  inflammation related to buildup of, 18–20, 21

role of, in circulatory system, 17–19
Chopped Asian Salad with Orange and Almonds, 163–64
circulation system
  cancer and, 37
  exercise and, 104–5
  massage and, 23
  yoga and, 25–26
Citrus Salsa, 150–51
cocoa, in diet, 93–94, 100
Coconut Aminos Dressing, 163–64
coffee
  diet with, 93, 100
  Parkinson's disease, 47, 93–94, 100
cognitive decline
  conditions needed for, 69–70
  diet and lifestyle for lowering risk for, 99
  glia cells and, 67
  lymphatics' role in, 72–73
  pathology of, 69–73
  people's fears about, 63
cognitive disorders. *See also* Alzheimer's disease; Parkinson's disease
  inflammation in, 73
colon, and digestion, 44
community, 138–39
coronary artery disease (CAD)
  diet and, 88
  exercise and, 108, 110
COVID-19
  immune system and, 34–36
  obesity and risk for, 36
cranial nerves, 45, 76
Creamy Hummus, 159
Crohn's disease, 49, 52
crossed-leg forward bend, in yoga, 190

Curry Bowl, 154–55
cytokines
    inflammatory process and, 8,
        35, 47
    lymphatic flow response to, 5
    lymph system clearance of, 9, 10

Dairy-Free Breakfast Smoothie,
    143
dark chocolate, in diet, 94, 100
deep relaxation, in yoga, 191–92
deep tissue massage, 128
dementia, 63, 74, 88, 102, 110
diabetes
    COVID-19 risk and, 35
    exercise and, 102, 105, 110
    glycemic index and, 92
    ketogenic diet and, 97
    lymph flow leakage in, 36
    Mediterranean diet and, 89,
        98, 99
    obesity and, 10, 85
    standard American diet and, 86
    sugar consumption in, 87–88
    yoga and, 102, 181, 187
diaphragm, and lymphatic system
        pressure, 4, 104
diet, 83–100, 140–76
    Alzheimer's disease prevention
        and, 72
    approaches to daily choices
        in, 94
    BDNF stimulation and, 80
    beverage pyramid in, 91
    brain health and, 47, 78–79, 81
    caloric energy and, 84
    chart summarizing best
        approach to, 89–90
    fats in, 86–87, 88, 90

fiber in, 87, 94, 99, 100
four main areas of concern in,
    84–86
high-carb approach to, 91–92
IBS and, 50–51
ketogenic, 97
lectin-free, 95
lymph flow and, 26–27, 79,
    89–90
Mediterranean, 80, 88, 92–93,
    95–97, 98, 99
mental ability and memory
    and, 77
micronutrient deficiency in, 88
mindfulness in eating and, 137
moderation in, 93–94
paleo, 97–98
personal variables affecting, 83
protein amount in, 87, 88, 90,
    97, 99
strategy in choosing, 84
sugar in, 87–88
support system additions to,
    100
types of approaches to, 94–99
vegetarian, 95
walking before eating and, 109
weight management and,
    59–60
whole versus processed foods
    in, 60, 85–86
digestive system. See
    gastrointestinal (GI) system
dinner recipes, 150–51, 168–75
disease. See also specific diseases
    conditions needed for chronic
        degeneration in, 69–70
    lymphatics' influence on, 7
dressing recipes
    with Apple and Shaved

Brussels Sprout Salad, 166–67

Avocado Dressing, 145–46

Coconut Aminos Dressing, 163–64

Lemon-Chive Dressing, 168–69

Easy Herb-Roasted Nut Mix, 176

eating habits. *See also* diet
  mindfulness in, 137

Eggs and Leafy Greens Greek Scramble, 162

endocrine system
  digestive system and, 48–49
  lymphatic system and, 2, 10
  relaxation in yoga and, 121, 125, 191–92

endothelial cells
  blood-brain barrier with, 68
  in lymphatic vessels, 6, 12

epigenetics, xviii–xix, 80, 107

exercise, 101–15
  BDNF stimulation and, 80
  benefits of, 101–2
  brain health and, 106–7
  cardio approach to, 110–11
  circulation improvement and, 104–5
  exhilaration after, 106–7
  factors in beginning, 108–9
  HIIT (high-intensity interval training) in, 113–14
  lymphatic flow increases in, 53, 78, 103–5
  memory improvement and, 107
  rebounder trampoline jumping in, 114–15

  strength training in, 111–12
  toxin clearance by, 104
  types of approaches to, 110–15

eye exercises, 183–84

fasting, in weight management, 60–61

fats, dietary, 86–87, 88, 90

fat storage
  anatomy of, 56–57
  health impact of, 57
  lymph and, 58

fiber, dietary, 47, 80, 87, 94, 99, 100

fish recipes
  Halibut in Parchment with Citrus, Green Beans, and Tomatoes, 174–75
  Roasted Wild Salmon with Citrus Salsa, 150–51
  Tuna Salad with Olives and Cucumber, 149

Five for Flow supplements
  for brain building, 177–80
  for fighting cancer, 37
  for fighting disease and inflammation, 8
  for gut health, 53
  for heart health, 27–28

flavonoids
  diet with, 27, 39, 42, 80, 94, 98, 100
  lymph flow and, 78–79
  new neuron growth and, 81

flow. *See* blood flow; lymph flow

foam cells, 19

food and eating habits. *See* diet

forward bends, in yoga, 188

freeze, fight, or flight response, 122

fruits, in diet, 89

gallbladder, and digestion, 44
gastrointestinal (GI) disorders,
   47–53
   leaky gut in Parkinson's disease
     and, 74–75
   lymphatic vessel problems and
     metabolic disorders in, 5–6
gastrointestinal (GI) system
   brain-gut connection in,
     44–47, 74
   celiac disease and, 95
   ecosystem components of, 41
   interaction of systems in, 48–49
   lymph flow and health of, 51–53
   obesity's influence in health
     of, 55
   osteopathic manipulation for, 51
   overall wellness related to,
     41–42
   overview of anatomy of, 42–47
   role of lymphatic system in, 10–11
genetics
   Alzheimer's disease and, 70
   cancer and, 30
   exercise modification of, 107
   heart disease and, 22
   lymph quality and flow and, xix
Ginger Vegetable Stir-Fry, 170–71
ginseng, 12, 80, 81
glia cells
   Alzheimer's disease and, 67
   function of, 67
   inflammatory response to
     brain toxins by, 75–76, 77
gluten
   lectin-free diet avoiding, 95
   toxic brain effect of, 79
glycemic index, 92
glymphatic system, 10, 73, 75–76
"good fats" diet, 99

grains
   in diet, 90
   lectin-free diet avoiding, 95
gratitude, 137
Guacamole, 158
Gundry, Dr. Steven, 96–97

half spinal twist, in yoga, 189–90
Halibut in Parchment with Citrus,
   Green Beans, and Tomatoes,
   174–75
Harvey, Dr. N. L., 58
HDL. *See* high-density
   lipoproteins
health. *See also* brain health; heart
   health
   obesity's influence in, 55, 59, 85–86
   weight gain and risk for
     problems in, 57
heart attacks
   causes of, 15, 21, 57
   exercise and, 102, 110, 111
   frequency of, 14
   lymphatics' role in, 105, 106
   saunas and lower risk for, 25
   spiritual aspects of our lives and, 131
heart disease, 13–28
   causes of, 15, 21, 22
   diet and, 86
   lymphatics' role in prevention
     of, 15–16
   proper and consistent
     maintenance for risk
     reduction for, 14–15
   shift in attitudes needed in
     approach to, 13–14
   treatment versus prevention in,
     14–15
   triglyceride levels in, 92, 109

heart health
  basic function of, 16
  deep breathing and, 24
  diet and, 26–27
  lifestyle choices affecting, 15–16
  lymphatic flow for
    strengthening of, 21–22
  massage and, 23
  meditation and, 26
  music and, 23
  obesity's influence in, 55
  osteopathic manipulation and,
    23–24
  overview of anatomy of, 16–18
  rebounding trampoline
    jumping and, 24–25
  saunas and, 25
  supplementation for, 27–28
  ways of improving lymphatic
    flow in, 21–27
  yoga and, 25–26
herbs
  for heart health, 26
  for lymph flow, 12, 80, 95
high-carb diets, 91–92
high-density lipoproteins (HDL)
  cholesterol and, 9, 18–19
  exercise and, 9, 102, 105
high-fiber diets, 100, 47, 87
high-protein, low-carb diet, 99
HIIT (high-intensity interval
    training), 113–14
hummus recipe, 159

imagery, in yoga, 192
immune system
  cancer response of, 32–34
  COVID-19 and, 34–36
  digestive system and, 48–49

IBS and, 50–51, 58
inflammatory response to, 32,
    50–51
innate and adaptive parts of, 34
lymph and spread of, 37–38
lymphatics' role in, 7–8, 21–22
lymph system transport in, 4,
    5, 19, 36, 37
osteopathic manipulation for,
    51
overview of inner workings of,
    30–33
inflammation
  Alzheimer's disease and, 71–72,
    73
  cardiovascular disease related
    to, 15, 18, 19–20, 21
  chronic degenerative disease
    and, 70
  immune response to pathogens
    and, 31–32, 50–51
  lymphatics' role in disease and,
    7–8
  lymphatics' role in prevention
    of, 15–16, 20, 21–22
inflammatory bowel disease (IBS),
    49–51
  common forms of, 49–50
  diet and, 50–51
  lymphatics and, 52, 58
insulin
  pancreas production of, 44
  sugar consumption and, 87
insulin response
  complex carbohydrates and, 82
  digestion and, 56–57
  fasting and, 61
  ketogenic diet and, 97
  sugar consumption and, 87–88
  yoga's impact on, 181, 187

intermittent fasting, in weight management, 60–61
interstitial fluid, 3

Japanese knotweed, 12
John of the Cross, Saint, 139
jumping, for lymph flow, 24–25

ketogenic diet, 97
Kipnis, Dr. Jonathan, 76

laughter, 137–38
LDL. *See* low-density lipoproteins
leaky gut syndrome, 50
lectin-free diet, 95
Lemon-Chive Dressing, 168–69
Lentil and Chickpea Stew with Turmeric, 165–66
Lewy body dementia, 74
liver, and digestion, 44
loneliness, 138–39
love, spiritual, 138
low-density lipoproteins (LDL)
    cholesterol and, 9, 18–19
    elevated triglycerides as risk factor for, 92
    exercise and, 102, 105
    fasting and, 61
lower back flexes, in yoga, 187
low-fat diet, 99
lunch recipes, 145–49, 163–67
lymph
    cancer spread and, 37–38
    factors affecting viscosity of, 3
    herbs for supporting, 12
    maintenance of body by, 3
    mechanics and biology of, 6–7

optimum diet for flow of, 83–100, 140–76
supplements for drainage of, 8, 177–80
three main functions of, 4–5
lymphatic massage, 128
lymphatic pump techniques, 23–24, 51
lymphatic system
    blood flow connected to, 22
    brain health and, 64, 72–73, 76–78
    cancer and, xiv, 29–30, 36–39
    cancer growth and sluggishness of, 38
    cardiovascular system and role of, 8, 9
    chronic inflammation resolution and, 20
    diaphragm's role in, 4, 104
    difficulty diagnosing problems in, 2
    disease fighting by, 7
    exercise improvements in, 101–15
    GI disorders and, 41–53
    GI system and role of, 10–11
    heart disease and, 13–28
    herbs supporting, 12
    immune cell transport by, 36, 37
    impact on systems of dysfunction of, 9–10
    interactions between other body flow systems and, 1, 2, 22
    lack of central pumping mechanism in, 4
    lack of measurement system for content in, xi, 11

lymph node removal's impact on, 39–40
maintenance of body by, 3
major systems affected by dysfunction of, 9–11
massage and, 127–29
mechanics and biology of, 5–7
neurological system and role of, 10
relationship of chronic disease, immunity, and, 7–8
secrecy of workings of, 2
spirituality and, 131–39
weight management and, 55–61
yoga for, 117–27, 181–96
lymphedema, and lymph node removal, 39–40
lymph flow
BDNF and, 79–80
components affecting, 6–7
deep breathing for, 4, 23, 24
diet for, 26–27, 79, 89–90
disease and slowness or stagnation of, xi, 2
exercise for increase in, 53, 78, 103–5
flavonoids and polyphenols for, 78–79
good health related to strength of, xi–xii, xvi–xvii, 1–2
gut health and, 51–53, 79
heart health and, 21–27
herbs supporting, 12, 26, 80, 95
impact of nutrient leakage in, 5–6, 8, 36
interactions between other body flow systems and, 1, 2
lack of central pumping mechanism for, 4
lack of measurement system for, xi, 11
massage for, 23
mechanics and biology of, 5–7
meditation for, 26
muscle contraction and, 103–4
music listening for, 23
obesity and, 10–11, 36, 57–59
osteopathic manipulation for, 23–24, 51
personal control over, 4, 7
saunas for, 25
as a secret river of health, xiii–xiv, xv–xvi, 1–2
sleep and, 79–80
spirituality and soul in, 132
supplements for, 27–28
toxins trapped and cleared away by, 3, 4, 7, 69–70
trampoline jumping for, 24–25, 78
ways of improving, 21–27, 78–80
yoga for, 25–26, 118–19, 125–26, 181–96
lymph nodes
action of cancer cells in, 37, 38–39
impact of removal of, 39–40
inflammatory bout in Crohn's disease and, 52
interstitial fluid flow through, 3
toxin clearance by, 3
white immune cells in, 5

macronutrients, 85
caloric energy and, 84

macronutrients (*cont.*)
    dietary need for, 85
    forms of, 56, 85
    ulcerative colitis and, 49–50
macrophages
    brain inflammation from, 75
    COVID-19 and, 36
    immune response and, 32
    role of, in circulatory system,
        9, 19
massage
    lymph flow and, 23, 127–29
    types of, 128
McKee, Dr. Dwight, xiii
McLanahan, Dr. Sandra, xiii
meditation
    how to start, 136
    lymph flow and, 26, 119
    spiritual practices using, 135,
        136
    yoga and, 195
Mediterranean Cauliflower Salad
    with Grilled Chicken,
    147–48
Mediterranean diet, 80, 88, 92–93,
    95–97, 98, 99
melatonin, in Alzheimer's
    treatment, 68
memory
    beta-amyloid in, 70–71
    diet's influence on, 77
    exercise and, 106, 107
    immune system and, 32, 34
    meditation and, 136
    neuron communication in
        brain and, 66
metabolic syndrome
    abdominal obesity and, 10
    diet and, 96, 97
    lymph flow leakage in, 59

microbiome, 47, 79, 94, 100
micronutrients, 85
    cancer prevention and, 30
    dietary need for, 85, 88
    dietary sources of, 27, 60, 94
    heart health and, 27
    processed foods and loss of, 60
    reasons for and impact of
        deficiencies of, 88
    supplements for loss of, 27, 96,
        177–80
    ulcerative colitis and, 49–50
mindfulness, 136–37
minerals
    dietary need for, 85, 88
    dietary sources of, 26, 89, 90
    processed foods and loss of, 60
    supplements for, 28, 96, 178
monosodium glutamate (MSG),
    71–72
mood, and exercise, 106
muffin recipe, 160–61
multiple sclerosis (MS), 51, 73, 77
music listening, for lymph flow, 23

National Cancer Institute, 86
neck extension pose, in yoga, 189
Nedergaard, Dr. Maiken, 76
nervous system
    digestive system and, 48–49
    lymphatic system and, 2, 10
    yoga's benefits for, 126–27
neurological disorders
    diet for, 92, 96, 98
    gut bacteria and, 79
    lymph-flow cleansing process
        in, 7, 77, 78
neurological system, role of
    lymphatics in, 10

neuron communication, in brain, 65–67, 74
nut recipes
   Easy Herb-Roasted Nut Mix, 176
   Spiced Walnuts, in Apple and Shaved Brussels Sprout Salad, 166–67
nutrients. *See also* macronutrients; micronutrients
   blood-brain barrier and, 68
   bodily fluids' transport of, xv, 6, 68
   impact of leakage of, 5–6, 8, 36
   lymphatic system's delivery of, xv
   massage for movement of, 23
nutrition. *See* diet

obesity
   COVID-19 risk and, 35
   exercise and, 102, 110, 112
   impact on health of, 55, 56, 59, 85–86
   lymph flow and, 10–11, 36, 57–59
   ways of fighting, 59–61
olive oil, in diet, 90, 92, 93, 96–97, 98, 99
osteopathic manipulation, for lymph flow, 23–24, 51

paleo diet, 97–98
pancreas, and digestion, 44
parasympathetic nervous system
   breathing and, 192–93
   GI system and, 46–47
   stress management and, 126–27

yoga's stimulation of, 124, 183, 186, 188, 190, 192
Parkinson's disease
   alpha-synuclein in, 73, 74, 76
   brain inflammation in, 73–74, 77
   diet and, 47, 93–94, 100
   lymph-flow cleansing process in, 76–77
   possible causes of, 46, 74–75
physical activity. *See* exercise
pleurisy root, 12
polyphenols
   anti-inflammatory effect of, 94
   diet with, 39, 42, 80, 94, 98, 100
   lymph flow and, 8, 27, 28, 37, 68, 78–79
   olive oil with, 90, 96–97
prayer, 137
processed foods, 60, 85–86
protein
   diet and, 87, 88, 90, 97, 99
   lymphatic transport of, 5, 10, 51, 59
pterostilbene, in Alzheimer's disease, 68

Quinoa Bowl with Apples, Ginger, and Cinnamon, 144

Randolph, Dr. Gwen, 52
rebounder trampoline jumping, 24–25, 114–15
recipes, 140–76
   for breakfast, 143–44, 160–62
   for dinner, 150–51, 168–75
   for lunch, 145–49, 163–67
   overview of diet with, 140–42

recipes (*cont.*)
  for snacks, 158–59, 176
  for week #1, 140–41, 143–59
  for week #2, 141–42, 160–76
  list of recipes
  Almond-Oat Lemon Blueberry
    Muffins, 160–61
  Apple and Shaved Brussels
    Sprout Salad with Spiced
    Walnuts, 166–67
  Black Bean Burger over Greens
    with Avocado Dressing,
    145–46
  Braised Chicken Tacos with
    Red Cabbage Slaw, 156–57
  Breakfast Quinoa Bowl
    with Apples, Ginger, and
    Cinnamon, 144
  Chopped Asian Salad with
    Orange and Almonds,
    163–64
  Creamy Hummus, 159
  Dairy-Free Breakfast Smoothie,
    143
  Easy Herb-Roasted Nut Mix, 176
  Eggs and Leafy Greens Greek
    Scramble, 162
  Ginger Vegetable Stir-Fry,
    170–71
  Guacamole, 158
  Halibut in Parchment with
    Citrus, Green Beans, and
    Tomatoes, 174–75
  Lentil and Chickpea Stew with
    Turmeric, 165–66
  Mediterranean Cauliflower
    Salad with Grilled Chicken,
    147–48
  Roasted Wild Salmon with
    Citrus Salsa, 150–51

  Rosemary Baked Tofu with
    Asparagus and Gremolata
    White Bean Mash, 172–73
  Spaghetti Squash with
    Pumpkin Seed Pesto and
    Shrimp, 152–53
  Spiced Chickpea Bowl with
    Roasted Tomatoes, 168–69
  Thai Curry Bowl, 154–55
  Tuna Salad with Olives and
    Cucumber, 149
Red Cabbage Slaw, 156–57
reflexology, 128
reiki, 128
relaxation, in yoga, 191–92
relaxation response, 124, 183, 188,
  192
resveratrol, in Alzheimer's disease,
  68
Roasted Nut Mix, 176
Roasted Wild Salmon with Citrus
  Salsa, 150–51
Rolfing, 128
Rosemary Baked Tofu with
  Asparagus and Gremolata
  White Bean Mash, 172–73

salads
  Apple and Shaved Brussels
    Sprout Salad with Spiced
    Walnuts, 166–67
  Black Bean Burger over Greens
    with Avocado Dressing,
    145–46
  Braised Chicken Tacos with
    Red Cabbage Slaw, 156–57
  Chopped Asian Salad with
    Orange and Almonds,
    163–64

Mediterranean Cauliflower Salad with Grilled Chicken, 147–48
Salmon, Roasted, with Citrus Salsa, 150–51
salsa recipe, 150–51
Satchidananda, Swami, 139
satiety hormones, in fasting, 61
saunas, for lymph flow, 25
serotonin, 44–45, 46, 138
shiatsu, 128
shoulder stand, in yoga, 188–89
Slaw, Red Cabbage, 156–57
sleep, and brain lymphatic flow, 79–80
small intestine, and digestion, 43–44
smoothie recipe, for breakfast, 143
snack recipes, 158–59, 176
soul. *See also* spirituality
    components of, 135–39
    stress management and, 133–34
    types of approaches in, 135–39
Spaghetti Squash with Pumpkin Seed Pesto and Shrimp, 152–53
Spiced Chickpea Bowl with Roasted Tomatoes, 168–69
spices
    for heart health, 26
    for lymph flow, 95
spirituality, 131–39. *See also* soul
    lymph flow and, 132
    types of approaches in, 135–39
spiritual love, 138
Stew, Lentil and Chickpea, with Turmeric, 165–66
stir-fry recipe, 170–71
stomach, and digestion, 43
strength training, 111–12

stress management
    benefits of, 126
    lymph flow and, 126
    soul and, 133–34
    yoga and, 121–24
sugar
    carbohydrates with, 91, 92
    in diet, 27, 87–88, 99
    fruits with, 94
    ketogenic diet and, 97
    processed food with, 85
    triglyceride levels related to, 92, 109
sulforaphane, 72
sun salutation, in yoga, 184–86
supplements for flow ("Five for Flow")
    for brain building, 177–80
    for fighting cancer, 37
    for fighting disease and inflammation, 8
    for gut health, 53
    for heart health, 27–28
    recommended list of, 177–80
Swedenborg, Emanuel, 132, 135
Swedish massage, 128
swelling, lymphatic pump treatment for, 24
Sydenham, Dr. Thomas, 137
synuclein, 74, 76

taco recipe, 156–57
tau protein, 71, 73–74
T cells
    COVID-19 depletion of, 36
    immune response to pathogens and, 32
Teresa, Mother, 138
Thai Curry Bowl, 154–55

Thai yoga massage, 128
therapeutic touch, 128
trampoline jumping, 24–25,
    114–15
triglyceride levels
    exercise and, 109
    heart disease and, 92, 109
    lymphatic vessel leaks of, 5,
        10, 36
Tuna Salad with Olives and
    Cucumber, 149

ulcerative colitis, 49–50, 53
upper back flex, in yoga, 186–87

vagus nerve, 45–47, 126–27
vascular system, and lymph flow, 4
vegetables, in diet, 89
vegetarian diet, 95
very low-fat diet, 99
visualization, 192
vitamin D, and schizophrenia,
    79
vitamins
    Crohn's disease and absorption
        of, 49
    dietary need for, 85, 88
    dietary sources of, 26–27, 89,
        90, 99
    immune system and, 34
    processed foods and loss of, 60
    supplements for, 27–28, 96, 177

weight gain
    health risk from, 57
    ways of fighting, 59–61

weight management, 55–61. *See
    also* diet
    diet as key driver in, 59–60
    fat storage and, 56–57
    intermittent fasting and, 60–61
    lymph flow's role in obesity
        and, 57–59
    Mediterranean diet and, 89
    ways of fighting weight gain in,
        59–61
    whole versus processed foods
        in, 60
White Bean Mash, 172–73
white blood cells
    cancer cell infiltration and, 38
    cardiovascular system and, 9
    exercise's impact on, 109
    inflammatory response and, 32
    intermittent fasting and
        availability of, 61
    lymphatic pump treatment for
        mobilization of, 24
    lymph system transport of, 5,
        37
    visualization and imagery for,
        192
Willett, Dr. Walter, 95
wine, in diet, 93, 97, 100

yoga, 117–27
    basics, 181–96
    benefits of, 181
    breathing exercises in, 119,
        181, 192–94
    5-minute quick lymph routine
        in, 196
    general cautions before
        beginning, 124–25

lymph flow and, 4, 25–26,
    118–19, 125–26
meditation in, 195
muscle activity in, 119–21
reasons for using, 118
relaxation response in, 124,
    188, 192

stress management using,
    121–24
20-minute lymph routine in,
    195–96

zonulin, 50, 51

About the Author

Gerald M. Lemole Sr., MD, received his medical degree and surgical residency at Temple University School of Medicine. He went on to do his cardiac surgical residency with Drs. DeBakey and Cooley in Houston, Texas. In 1968, he was a member of the surgical team that performed the first successful heart transplant in the United States. He returned to Temple University as chief of cardiovascular surgery. In 1986, he was appointed chief of cardiac surgery at the Medical Center of Delaware, where he performed the first coronary artery bypass in Delaware. Upon stepping down as chief of cardiac surgery, Dr. Lemole became certified by the American Board of Integrative Medicine and became director of Integrative Care at Christiana Care.

He has written more than 150 medical articles, book chapters, and editorials for professional publications. He has written five books on integrative medicine and has lectured and operated all over the world. His books include *The Healing Diet* and *Facing Facial Pain*.